CONSERVATION GARDENING AND FARMING® SERIES
ERIES C: REPRINTS • Edited by Bargyla and Gylver Rateaver

Organic SMALL FARMING

by Hugh Corley

THE EXCITING STORY OF SCIENTIFICALLY CONTROLLED METHODS USED ON PUCKETTY FARM

2nd Edition Published 1975
by Bargyla and Gylver Rateaver, Pauma Valley, Calif. 92061

Cover Design and Title Page by Jon Gnagy

SECOND EDITION

Published by Bargyla and Gylver Rateaver
as a reprint in the
Conservation Gardening and Farming® Series
Series C: Reprints
ISBN 0-9600698-4-4
Library of Congress Catalog Number 74-33122
Cover art of this second edition by Jon Gnagy

First edition published 1957 by Faber and Faber Ltd.

COVER DESIGN AND TITLE PAGE BY JON GNAGY

Published in the U.S.A.
by the editors
Bargyla and Gylver Rateaver
Pauma Valley, Calif. 92061

This edition

is dedicated to

JHVH,

the Creator,

whose Garden of Eden included no

chemical fertilizers or poison sprays

To B.

as a beginning

NOTE

The American reader will have to read "mixed cereals" for "*dredge corn*"; "temporary pasture" for "*ley*"; "alfalfa" for "*lucerne*"; "grain" for "*corn*"; "fertilizing" for "*manuring*"; (FYM is what we call manure — farmyard manure); "manure" for "*dung*"; "calcium nitrate" for "*nitrochalk*"; "corn" for "*maize*"; "plantain" for "*ribgrass*".

A fallow is land left without cropping for a season or more. A bastard fallow is the same but with a green manure crop grown on it, to be turned under.

Haughley is the experimental farm which was owned by the Soil Assn. in England, but which is now under the Pye Trust. The current address for both is: Walnut Manor, Haughley, Stowmarket, Suffolk, England.

The periodical of the Soil Assn. was called *Mother Earth,* but is now called simply: *Soil Assn.*

Newman Turner began a publication called "*The Farmer,*" but it went out of print at his death and is not available.

Lodging of cereals refers to the way the plants flatten under the impact of wind, rain, hail or heavy snow, due to lack of strength in the stems, sometimes from potassium deficiency.

Cwt is 112 lbs. in England, 100 lbs. in the U.S.

Score is a British weight used for pigs, and is about 20 lbs.

CONTENTS

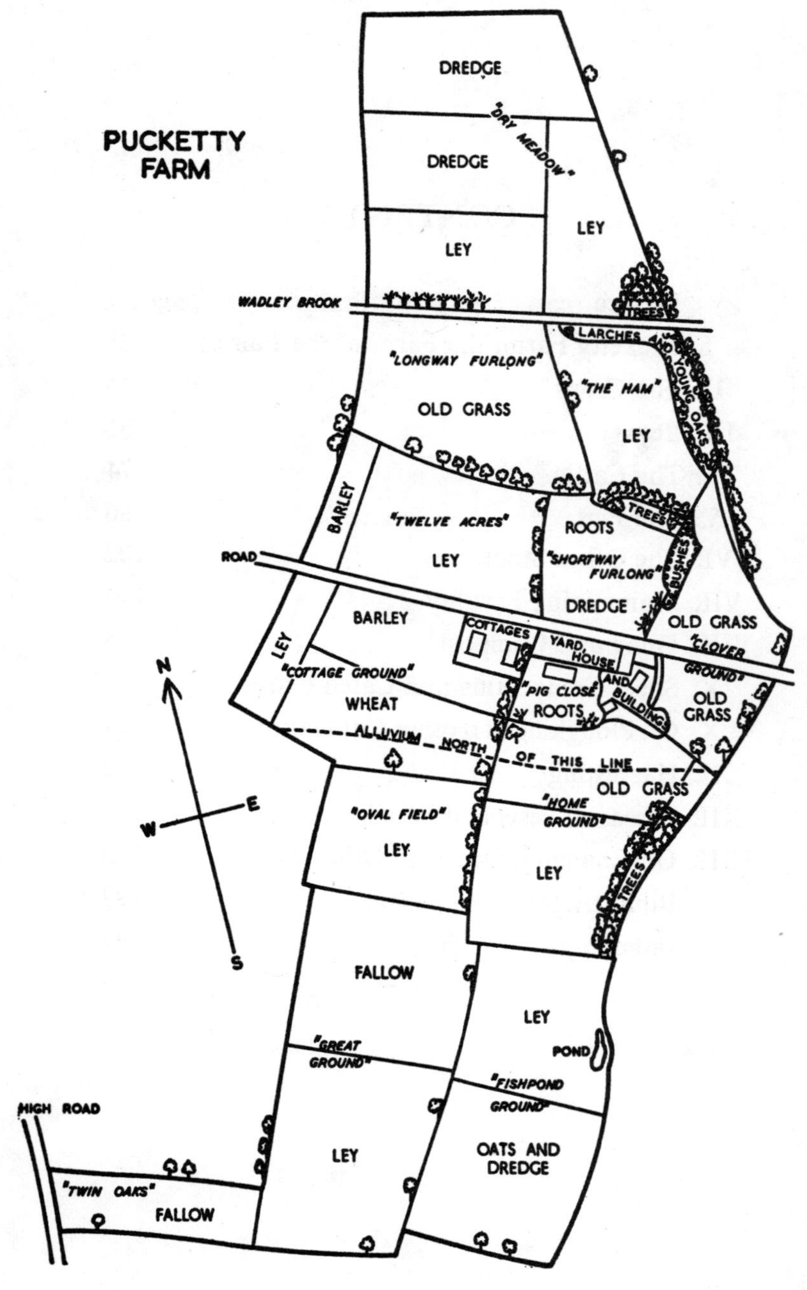
PUCKETTY FARM
DREDGE
"DRY MEADOW"
DREDGE
LEY
LEY
WADLEY BROOK
TREES
LARCHES AND YOUNG OAKS
"LONGWAY FURLONG"
OLD GRASS
"THE HAM"
LEY
BARLEY
"TWELVE ACRES"
LEY
TREES
ROOTS
"SHORTWAY FURLONG"
BUSHES
ROAD
DREDGE
OLD GRASS
BARLEY
COTTAGES
YARD
HOUSE
"CLOVER GROUND"
LEY
"COTTAGE GROUND"
WHEAT
"PIG CLOSE"
ROOTS
AND BUILDINGS
OLD GRASS
ALLUVIUM NORTH OF THIS LINE
N
OLD GRASS
E
W
"OVAL FIELD"
LEY
"HOME GROUND"
LEY
TREES
S
FALLOW
LEY
POND
"GREAT GROUND"
"FISHPOND GROUND"
HIGH ROAD
LEY
OATS AND DREDGE
"TWIN OAKS"
FALLOW

ILLUSTRATIONS

PREFACE

I am one of many organic farmers in Britain. The reader would be wrong to suppose that I am one of the cleverest (and that is not false modesty). I am a most undistinguished farmer except in one respect. The thing that distinguishes me from most, but not all, of my fellow organic farmers is that I have been willing and able to find time to write a book about my farming. Remember that it is not, on that account, the only or even the best kind of organic farming. Even orthodox farmers do not all use the same methods, and this is even truer of organic farmers where the flexibility of natural processes, the diversities of soil and climate and the experimentalism which necessarily goes with a young movement have resulted in almost as many ways of farming organically as there are organic farmers. There is no need to apologize for this, for in this variety of farming techniques lies the road of progress.

I therefore trust that any criticism, implied or otherwise, of some of these other methods will not be taken as an attack on what is sometimes contemptuously called 'Muck and Magic'. For these other organic farmers are what I can only call my colleagues in a most important work, the wooing of farming England back to sanity.

I hope this book will not strike the reader as being either arrogant or complacent. I spoke of a new movement. There is nothing new about farming organically, and no grounds for boasting about it, much less for complacency, because we have found out a little about it. Man has probably been farming for ten thousand years. And it is only in the last century that it has become possible to base a civilization on the precarious and ever shifting foundations of chemical farming. The only new factor is the way in which biology is coming to the rescue of farming which has been so hopelessly led astray by chemistry with its abysmal and apparently incurable ignorance of all things biological. Of course there is the newish science of biochemistry. But, at least as far as farming is concerned, it has shown more inclination to treat biology as a branch of chemistry than to combine the chemist's knowledge with the biologist's understanding. It is from the biologists and the biologically-minded farmers that the new impulse has come which has made a new movement of an ancient craft.

This new movement has, of necessity, something of the crusade about it. And the danger of a crusade is that it may so inspire its adherents with enthusiasm that their minds lose that scientific impartiality which is so necessary if we are ever to make a case that will convince the orthodox of the error of their ways. I have always recognized this danger and tried to guard against it in myself, and if the reader may sometimes find me over-cautious in my claims, he will, I hope, agree that the book is down to honest earth, that it deals, in fact, more with muck than magic.

Some of the ablest brains in farming (and that means in the whole country) are engaged in this movement to return to sound, organic methods, but we need more organic farmers. Many important discoveries in human history have been made by chance. In the days of the Stone Age farmers, progress must have been very slow not only because so little had been discovered, but because there were so few men to make discoveries. Much the same applies to-day. Organic farmers are still in a minority in the countries of western civilization, and a mere increase in their numbers would inevitably help (and is helping) the movement. The fact that organic farming is making rapid progress despite a relative shortage of organic farmers, and despite lack of the vast financial resources of orthodox or chemical farming for research and propaganda, is proof of its essential soundness.

I thought of calling this work 'Biological Farming', but decided against it. Much as farming owes to science, and to biology in particular, true farming will always remain an art.

I have had three main types of reader in mind while writing. One is the beginner at farming who wants to know all he can find out about organic methods, even the most elementary practical details. One is the non-farmer who is nevertheless interested in farming, perhaps because he hopes to farm, has friends who do, is a keen gardener or merely takes an interest in the way his food is grown, as he has every right to do. The third type (and he must be patient with explanations that are not needed for him) is the ordinary, sensible farmer who has been led into chemical farming rather against his better judgement, swayed by the weight of propaganda or the threat of being called out-of-date and inefficient, and who is wondering if all is really well with orthodoxy. Among some who wonder thus are at least some of the country's official agricultural advisers.

To the wavering orthodox farmer I would say: take the plunge. Put aside a small part of your farm and run it on organic lines and see for yourself what happens. I ask no more than that. It is not pig-headedness that has converted me from wavering orthodoxy to unwavering opposition. I have been repeatedly re-convinced of the general soundness of organic ideas by my observations of nature on the farm. And if this book seems rather a hotch-potch of untidied-up observations it is because I am not blessed by possessing an all-explaining philosophy of nature and her ways. I can only record what I know, and guess here and there at possible explanations. The organic fanatic who is hoping for a sort of divine revelation of truth that will tie up the odd ends and confound the philosophies of chemical heathendom has opened the wrong book.

There may be some confusion in the reader's mind as to what is meant by 'this year' in the text. The fact is that I have been writing and revising the book intermittently for several years, and many months must now elapse before it can be published.

On the subject of crop yields in 1954, potatoes were down to about 6½ tons per acre. This was due to a number of causes including late planting and a bad season. It is not a fair indication of what you may expect from good land organically farmed and where potatoes form an important crop. Fodder beet yielded 24 tons per acre on the best plot, and Red Intermediate mangolds 29 tons. Allowing for the edible tops of the beet, that crop outyielded the mangolds in mere tonnage, without allowing for the higher sugar content of the beet.

In 1955 I ploughed down 23 tons of dung per acre and worked about 20 more into the surface soil for fodder beet. I hoped for great things. The beet were sown in good time and came up well in early May. Late in May frost killed the leaves. They came again and were again killed by frost on June 10th, only to be repeatedly waterlogged after that till nearly the end of the month when there began the tremendous drought which has not ended properly at the time of writing. Under these conditions the beet could hardly be expected to survive, let alone thrive. They did survive and have yielded about 10 tons (of roots only) per acre. On two farms, at least, in this district, which are run by excellent orthodox farmers, the roots were a total failure this year. If I do not grow a bumper crop of beet next year I shall think seriously of giving up this crop which is so excellent when it does well. However the dung used is not wasted, as artificials would largely have been wasted. It will mean good crops for several years following the beet.

In making my acknowledgements pride of place must go to Mr. David Spreckley of the Landsman's Library, Hertford, Huntingdon. Not only has he written much of the bibliography (and I am sure will be delighted to supply those and many other books) but it was in response to his suggestion that I originally decided to write this book.

If the book has virtue, it is largely due to Mr. Lawrence D. Hills, who has nursed and shaped, and made suggestions from the beginning.

To Mrs. K. M. Studholme go the laurels for translating my frenzied, manuscript scrawl into legible type. And I have to thank my brother, Mr. Michael Corley, for the beautiful photographs of hand-stacking wheat, the silo, and the crop of autumn-sown Atle wheat, and for help with the proofs.

Most of the other pictures were taken by myself with members of my family acting as measuring rods. These pictures lack the professional touch, but they do show, I think clearly enough, that excellent crops can be grown by organic methods. For none of the crops shown owe anything whatever to artificial nitrogen.

I must also thank my mother and my wife for much encouragement and for correcting slips of punctuation and so forth, my eldest son for inscribing the map of the farm, and my mother again for reading the printed proofs and doing most of the work for the index.

H. V. CORLEY

Pucketty Farm,
17th Oct. 1955.

CHAPTER I

PUCKETTY FARM

(THE FARM OF THE FAIRIES)

Here is a farm of 185 acres in the Upper Thames Valley. It is a very ordinary farm, arable and dairy mostly but with other stock as well. The only peculiar thing about it is the farmer, who, by the standards of the majority, is a crank.

It is a fairly compact farm, two or three fields wide, running more or less north and south with 70 to 80 acres at the southern end on heavy Oxford clay and the rest on alluvial soils over river gravel. The clay soils are extremely heavy, almost pure clays with a substratum of Oxford clay 350 to 400 ft. thick, absolutely impervious to water. The alluvial soils vary from very heavy clay loams to soils which are relatively light. And they vary in usefulness from really excellent soil to patches of very poor stuff where even heavy manuring produces only moderate yields. And the change from good to bad may take place in only a few yards.

But on the whole the land is good and I do not grumble. They say it is only a poor workman who blames his tools.

Men have been living at Pucketty, or, at least, hunting over it since time immemorial. I have found flint knives and arrow-heads on the land, and in 1953 one of my men hoed up a Roman denarius, a penny, of the Emperor Diocletian, in my fodder beet.

The ridge and furrow of the heavy clay land runs right through field and even boundary hedges and on up the side of the hill, showing that the land was farmed on the open-field system in feudal times.

Not all the men who have farmed Pucketty have been good farmers. And not all of them are to blame for that. Some were the victims of bad times when the cry for cheap food led England to neglect her own farm land in favour of a sort of spiv existence of smart trading; the days when gold was thought to make a people great instead of the food they ate and the thoughts they thought and the deeds they did.

But the land still survives. As I hope to show, it is, most of it, in pretty good heart now and, with God's help, it will continue to improve. For one thing I am profoundly thankful. I became responsible for this little bit of England before artificials were as widely used as they now are, and before the chemists (finally bidding farewell to reason and humanity) invented a host of deadly poisons with which to simplify such farming jobs as weeding crops and cleaning land.

Let me describe the farm as it was in the summer of 1953. It was a good summer. I seldom grumble about the weather, being thankful to my ancestors who chose this land and climate and fought for it, and to Almighty God who manages the weather much better than I should do, even were I able to affect it.

Then let us start at the southern end of the farm. Here is a small field, 5 acres, of heavy clay soil with a slight northern slope. In winter it is apt to lie cold and wet, but it is good land, and in summer has proved its worth with good crops of oats, wheat, clover, beans and even potatoes when recently ploughed from old turf during the war. This year it is filthy with all sorts of weeds and really needs more cleaning than I have been able to give it. For, let me assure the reader, this is not a perfectly run farm. Things do not always go according to plan, and I will try to describe the farm as it really was, this summer, and not as I wish it had been.

Next, Great Ground, is a ley. It is a herbal ley of deep-rooting plants and has been and still is a useful pasture. It is late to start growth in the spring owing to the clay soil and the slight northern slope, but it is not as wet as some of my heavy-land fields, and has grown good arable crops in its turn. It has also grown, in the early days of the war, the finest heads of wheat I have ever seen. The crop was a light one, the plants having been thinned out, on old, ploughed-up, worn-out grassland by wireworms and rooks. As so often happens with thin crops of wheat the individual plants were very fine, each producing a considerable number of very large heads.

Some of these wheat heads I have preserved in bottling jars. And one, at least, has a spikelet with 8 grains in it. Seven in a spikelet, almost unheard of in wheat, is common among these great ears. A spikelet is a section of an ear of wheat or other 'grass'. It is the thing you pull off a head of ryegrass when you play 'She loves me, she loves me not. . . .' There are commonly 10 or 12 of these on each side of a good head of wheat. In these heads of mine there are 13 to 15 spikelets on each side and they are so crowded on the stalk that the head twists near the apex in order to fit them in. The largest spikelets are always in the middle of the ear. One of these ears contained 123 grains of wheat. Others may well contain more. Growing thinly spaced in the field the plants developed straw like reeds and were well able to support these big ears. The variety was Wilma.

Only this year, 1953, have I grown heads anywhere near as good, not quite so good even now, and then only on the edge of the field. But this time I had the satisfaction of seeing them packed close together in a heavy crop of wheat.

But to continue our tour. The other half of Great Ground lies flatter and wetter. This year (1953) I have tried the experiment of mole-draining it. Two tiled mains, duly shingled, have been laid across the wettest and lowest parts of the field, and we used two tractors to pull a small mole plough (made by Tamkins of Essex) up and down the field. As a result I have risked putting in half the field to wheat. It was fallowed from necessity, because the mole-draining operations went on into the summer, and weeds got out of hand before it could be levelled again and ploughed.

On the east side of Great Ground are the two halves of Fish Pond Ground. The upper, southern, half again lies drier. This year 8 acres of it were in winter oats. They were a thistly crop, but not too bad seeing that they were sown very late last year and got very wet even before the corn was up through the ground. The straw was, much of it, 5 ft. high, but only a little was lodged, under trees and beside the hedge. The remaining 2 acres, always wet in winter, were left and sown with dredge corn (oats, barley and peas) in the spring. The wood-pigeons, as usual, ate most of the pea leaves for weeks, thereby reducing their contribution to negligible proportions, but a nice crop of dredge corn came to harvest all the same, though the catchy weather in September made harvest a little difficult.

The other half of Fish Pond Ground is in long ley. Again a deep-rooting, herbal mixture. But it has been down long enough. The chicory has dwindled, and it is time for the plough to come in.

Over the hedge to the north is what I call Home Ground. The greater part is in Italian ryegrass and trefoil. As usual I had intended to plough this early in the summer and put it into kale or some such fodder crop. But as usual I didn't. For one thing we were too busy with other work and for another this catch crop, sown under corn last year, was far too useful. So it stayed on from being a catch crop and was ploughed in the later summer for winter corn. By that time the ryegrass had given up the struggle, but the trefoil was doing manfully, and will help the winter oats with the nitrogen of its root nodules and the humus from roots and leaves together.

This is the first year I have dared to put this wet, flat field into winter corn. Some years it has been too wet in May to drill corn. But this time we have mole-drained it straight into the ditch. It may not be such a good job as with tiled mains, but it is quicker and cheaper and should be effective for at least one winter and much better than surface water furrows.

About 6 acres of Home Ground is in old grass. It has probably been in grass since Napoleon's day. This year I shut it up for hay. The weather did not admit of hay-making until about the 22nd June. And weeds grew alarmingly in all our green crops, so we got all behind with hay-making, as did most farmers. The other hay crops were ripening off fast and had to be cut to stop them spoiling. In the end Home Ground never was cut. Its indigenous strains of grass were still comparatively leafy in late summer, and we have had to graze, first with the cows, and soon now the young cattle will go in there and get a living from it for at least the first part of the winter.

Next summer it will want tight, close grazing to restore the clovers in the sward, and the mower will have to go in at least once, and more often if possible, to knock down those patches of creeping thistle which have flourished appallingly in the absence of any mowing this year. But the long rest from mowing and grazing will mean early spring growth next year, for the grasses have all built up good food reserves in their roots and leaf bases. We shall get an early bite as surely as if I were to dose the field with artificial nitrogen in the spring. Part of this piece is on clay and part over gravel.

West from Home Ground, behind that line of elms and oaks, is Oval Field, another field of ridge and furrow and heavy loam. But not *quite* so heavy as the others. It is a field whose fertility I have built up without the use of the dung cart.

Years ago I gave it 6 cwt. (112 lbs. = 1 cwt.) to the acre of compound fish manure and a coat of lime. I believe chalk and slag would have done as well or nearly as well as a beginning.

After one or two corn crops I sowed a fairly orthodox ley. It was not a great success, and during its second winter I stocked the field with in-calf heifers. They got some hay and some roots or kale and a great deal of straw to eat, and they trod the field and the straw they didn't eat into a mess. The next summer I fallowed it and sowed mustard. The dung and urine of the cattle, the fallowing and the remains of the clover supplied plenty of nitrogen to rot the straw and grow a tremendous crop of mustard. Some of it was setting seed and all was in flower when eventually we ploughed it in.

The following wheat crop was disappointing—about 18 cwt. to the acre—probably due to insufficient consolidation after the stalky mustard had gone in.

Following wheat I sowed oats and vetches for silage. The oats grew up to 6 ft. high and the vetches were a solid mass 4 ft. high in between the oats. Experienced farmers declared they had never seen anything like it. What the yield per acre was cannot guess. But the oats were ripening long before we could finish making silage.

The plough followed the mowing machine fairly closely as we worked our way across the field, and another crop of mustard, sown in August, followed the plough and cultivator. I also experimented with disking the stubble instead of ploughing it, but of that more later.

The mustard and about 3 cwt. per acre of rock phosphate were ploughed in and the field left rough for the frost. But there was no frost, nothing but rain and more rain. In spring there was no tilth and I had to fallow to try and make one. But drought during the summer prevented us making a fine tilth on the heavy soil until July when rain came and the soil 'slaked' into a lovely soft seed bed. I sowed a ley and late into November that year the grass was uniformly bright green all over the field, being no richer colour where dung and urine had fallen—a sure sign of abundant fertility. The field is still in ley now.

Working north again we come to what I have named Cottage Ground. The western 2¾ acres is wet and undrained and down to 'permanent' grass. I may be able to improve its drainage with the subsoiler, now that I have one, for we are now over the gravel, and drainage is at least possible through that.

The rest of the field grew 10 acres of wheat this year and 6½ acres of late sown barley. The wheat, following dunged Italian ryegrass, was a lovely crop. The barley, sown just before the Coronation, was a very useful crop considering its lateness. It will make a lot of pig food.

That part of the field was foul with squitch (couch) after two corn crops following inadequately hoed kale. The first corn crop was a nice piece of dredge corn. The second a tremendous crop of S.147 winter oats whose straw, over 5 ft. tall in many places, supported magnificent heads of grain. We got it cut just in time before a thunderstorm!

But the oats left the ground foul despite their density, for once they were gone the squitch romped away. So during April and May we worked the soil with a cultivator and drag harrows till almost all the squitch was lying dead on top of the ground.

I decided to chance a third corn crop and sowed Earl barley—chosen for its early ripening. We drilled both ways and harrowed in with the drags, for the dead squitch would allow of no smaller harrows being used.

The result was a pretty piece of barley, too pretty at one time. The lush growth in June caught an attack of mildew (the seed was bought) and I despaired of the crop. But I don't believe in worrying. Things have a way of turning out better than we expect, and sure enough the mildew checked the growth of the crop enough to stop it lodging. The corn then recovered and grew away to a reasonable height and all was well, only the grain was not quite well enough filled to make a brewer's sample. The yield was about 14 cwt. per acre. Not bad for such a late-sown crop.

Italian ryegrass, trefoil and a little white clover are coming on in the stubble to give my cows an early bite next February or March—I hope. The dead squitch, meantime, is rotting or has rotted back into the soil making excellent humus. Some which came up in the drags we picked up and put into a 'cock' in the temporary rick yard at the edge of the field by the gate, and there it stands, dead as doornails, waiting for a use as bedding or in compost heaps. None was burned.

Cottage Ground was wet and badly drained when I came to the farm. But it overlies gravel, at least for most of its extent. I sowed catch crops and leys of red clover and leys containing cocksfoot. The roots of these, together with the increasing worm population (also encouraged by dunging the field) and drainage made the soil porous down to the underlying gravel, and throwing out the ditch at the end of the field allowed the water to move sideways from the gravel into the ditch and thence ultimately to the river Thames.

More recently I have pulled the mole plough through the field. Standing on the 'plough' enables you to feel when the mole is going through clay and when through gravel. The water will travel through the gravel patches anyway. And the mole through the clay joins up one gravel patch with its neighbours with very good results.

The drainage of the gravel land is a two-edged sword. Sometimes in wet weather rain elsewhere will cause the water table in the Thames Valley to rise and a field will flood from below, the water rising through the worm and root holes and coming out on the surface of the field.

We have now travelled along the farm till the farmhouse and buildings are due east of us. But between them and Cottage Ground is Pig Close, a little paddock, which was never mown, never ploughed. At last its fertility became colossal with endless grazing and feeding of stock there in the winter, and when it got badly trodden in a wet winter I decided to plough.

Thousand-head kale was a fine crop, despite drought, in 1952, and this year another crop of thousand-head kale on part of the fields has grown great plants that came above my shoulder (I am about 5 ft. 11 in.) and that, I must admit, in unaided competition with weeds. For the fodder beet kept us too busy (until haymaking made us even busier) to allow much time for cleaning the kale.

The other part of Pig Close this year has grown a fine crop of Yellow Daeno fodder beet, with roots from which, incidentally, we took first prize at our local Ploughing Society's Show.

We are now at the widest part of the farm and are on the alluvial soils, having left the clay behind us—or rather below us. For it runs in a great sheet below the alluvial soils of this part of the Thames Valley, and below the river itself. The gravel and soil were laid on top of this clay during the Ice Age and just after it, and you only have to dig through a few inches or a few feet of gravel below the subsoil to reach the familiar blue clay which underlies the subsoil directly at the south end of the farm.

North of Cottage Ground and across the road is 'The Twelve Acres'. Two acres of this were barley this year—a thistly crop where cleaning next summer will be essential. The rest is in lucerne-grass-clover ley. It cut a good crop of hay this year and we have grazed the aftermath, being too busy to do anything else with it, even if I had wanted to. The hay is stacked in bales in the corner of the field on the high ground.

East of Twelve Acres is Shortway Furlong. The road to the farm crosses this field. To the south of the road is fodder beet, Pajberg Rex X, a gappy crop (which got us second prize at the Show)—only 20 tons or so to the acre—patched here and there with cabbage plants and carrots—the latter at any rate for our own use. A better plan is to sow a little kale seed mixed with the fodder beet seed. Then if the beet fails in any place a kale plant can be left. I did this in 1954.

Then come, north of the road, 2½ acres of dredge corn, a heavy crop, which lodged in one or two places. Then swedes, potatoes, more fodder beet and the rest of the field thousand-head kale. This crop did receive a little help in its fight with the weeds. It is a fine crop. There might have been more leaf and less stem if we could have singled the plants to a yard apart. But with most of the crop 5 ft. or more in height and a lovely healthy green colour there is really nothing to complain of. The last field the road crosses is the one in the corner of which stand the house and garden and the milking parlour. It is a field of permanent grass, down for scores of years at least, and full, among other herbs, of dandelions. A ley might produce more food per acre, but not food the cows would enjoy more. They are always glad to come back to 'Clover Ground'.

A word in defence of old pastures. They stand treading in winter better than most leys. They contain a much better variety of herbs and different grases than even a herbal ley usually does. And they often get scant attention.

The ley will get residual fertility from arable manuring and then special manuring (often very heavy) for the ley itself. It is rested, electrically fenced, rotationally grazed, top dressed, harrowed, rolled, cosseted and fussed over and, hardly surprisingly, produces more than old grass.

The old grass is often neglected. It is not manured. It is grazed at the wrong times, rested at the wrong times, mown too often or never at all. No wonder it is less productive than a ley. Nevertheless many vets hate leys and favour old grass both for hay and grazing. Certainly no ley can produce such lovely hay for calves and sick animals as a fertile old meadow.

Clover Ground and Shortway Furlong both have lovely loamy soil which is light (by the standard of most of the farm) and that in Shortway Furlong is easy to work. I mean to keep it so.

The farm now narrows towards the north. On the west, lying beside 'Twelve Acres' is a similar shaped field, 'Longway Furlong'. A furlong was a furrow long—the length a team of oxen could plough before halting for a rest. So here is another glimpse of farm history for us. Did the oxen wear those funny, little, broad, flat iron shoes we so often plough up? I don't know.

Longway Furlong is the best permanent grassfield I know for a heavy cut of hay. In 1943, 1945 and 1952 it produced truly astonishing crops, with another good but not exceptional crop this year, the two stacks of bales, one at each end of the field. Last year (1952) I also baled the crop, and I took the trouble to weigh some of the bales, count the others and estimate the yield. It came to 3 tons of hay per acre. As the average for *leys* for the whole country is 30 cwt. and for old grass 18 cwt., the reader will see that this is no ordinary field. I claim little credit for this fertility. I have maintained and even improved it; but it is a naturally productive field. However a noticeable improvement has taken place in the herbage—thanks largely, I believe, to dung and occasionally slagging and grazing for a whole summer instead of mowing. In 1955 hay from this field, made by pick-up bale, took first prize at our district Ploughing Society's Show. I only mention this small success to prove that organic methods, as described in this book, can produce and maintain a herbage second to none produced by other methods.

Beside the end of this field is 'The Ham', an old word said to mean a pasture that was not mown. It is surrounded by larch trees and dense bushes on the north-east and south sides and a fair hedge on the west. It is in fact a bad field to make hay in, but a wonderfully sheltered field in winter. It is now in its third year of ley, and a mass of white clover after early mowing for hay followed by hard grazing. I expect to stock it severely with cattle this next winter, dung it and plough in time to sow my roots there next spring. So the stack of hay bales is securely fenced with barbed wire against cattle.

North again, over the Wadley Brook (which drains all the land hereabouts into the Thames) is a single large field, 'Dry Meadow', of 32 acres.

When I came to the farm its herbage was mainly wild carrot and moon-daisies and it took it from January to October to grow a moderate hay crop.

It is now divided into four more or less equal parts. On the left, running beside the brook, are 8 acres which grew an emergency hay crop of Italian ryegrass and trefoil. This was sown as a catch crop and left for hay when the ley failed on the next piece. Part of this first piece is very good heavy land. It was sown with mustard early in September. But conditions were rather too dry and too late and there is only a poor crop.

On this next piece, the middle plot, I sowed my own seed of dredge corn and it made a picture, being pretty clean as well as a good crop, for we did plentiful cultivations before sowing. The land here is not good. Twenty-four tons of dung to the acre and a bastard fallow and a crop of mustard ploughed-in in 1951 could only produce 22½ cwt. of wheat to the acre in 1952. It is now undersown with a red clover-ryegrass ley for hay and seed in 1954.

The top piece of Dry Meadow, the furthest bit of the field from the farm (and the dung cart), is growing a self-sown clover ley at the time of writing which volunteered in early sown dredge corn. It is not surprising that it came, for in 1952 I left the aftermath of a fairish ley and ploughed it in when it was seeding. This has produced masses of humus and a worm population which for size and numbers is almost alarming.

I was a bit too clever sowing this lot of dredge corn. The exceptionally dry spring went to my head and I sowed part of the field at the end of February. The result was thinning of the crop by rooks and pigeons and a mass of charlock which germinated when the corn was sown and we could do no more harrowing.

I will not spray my charlock and poison everything from my

bees to the man who sows the stuff, so I mowed the affected parts of the field. My tractor driver managed to set the knife above the thickening in the corn stems where the young ears were forming, but low enough to decapitate most of the charlock flowers. The result was an awful mess for a few days and then a crop which really looked as well as the parts which were not mown.

The triangular piece of land in the corner next to the little group of limes and ash trees helped to grow the two stacks of bales by the gate. Actually there must have been 3 tons of potential hay to the acre at one time, but it had 'shrunk' a bit before we could cut it this year. Even so we got a lot of hay off it and it had been grazed till the beginning of May. The aftermath was ploughed in while still young and green, and we hope to sow winter dredge there.

That then is the farm. I have not mentioned all the trees (rapidly becoming curiosities in this part of the world) or the ponds, or the overgrown, neglected hedges. Yes, they are, mostly, neglected. They had been neglected for 50 years or more when I took over, just before the war, and building soil fertility by organic means takes so much time and hard work that I have continued to neglect the hedges.

I know one farmer who looks after his hedges in a most business-like and excellent way. But he burns everything from hedge trimmings and straw stacks to dung heaps and poor hay. It would seem that farm life is too hectic nowadays for adequate dung carting and adequate hedge trimming, and especially is this so on a farm like mine which has far more than its fair share of ditches to be looked after, most of which drain other people's land!

One thing we should have remarked on in Longway Furlong is the windmill water pump and water tower, supplying the farmhouse, cottages, dairy and drinking troughs in most of the fields. In very dry summers we go a bit short of water even now, shorter still since the local authority, at enormous expense, set up a pumping station higher up the Thames Valley. This takes water from the Thames to supply a number of surrounding

1. Silo empty.

2. Silo full.

3. Poppy, aged $13\frac{1}{4}$ years. She has averaged $10{,}384\frac{1}{2}$ lb. milk with her first nine lactations, counting only the official 305 days in each case. She looks like giving another 1,000 gallons in her present, tenth lactation.

4. Tulip, aged 10 years, has given 10,121 lb. mostly on three teats after an accident that tore off one teat. She suffered almost no pain from this ghastly wound, and the injured quarter stayed healthy.

villages, all of which already had excellent water supplies of their own. Incidentally this lowering of the water table adversely affects yields of crops in time of drought. But shortages apart, the improvement on ponds, both in efficiency of management and in health of cattle, is wonderful. Johne's disease has become a very rare visitor since the water-troughs came.

I have a windmill on a pylon 30 or 35 ft. high. There is another pump which we can work with a portable petrol engine if need be, but it is rare for us to need to do so. The exception to this is when there is a shortage of water. Then the wind may blow when there is no water to pump, and later there may be some water but no wind. But on the whole the wind is generous with its power, and we often stop the mill in winter, spring, and wet summers and autumns because the storage tanks are overflowing.

Electricity might pump with less trouble still, but it would have had to be taken there on poles which would have been a nuisance and an expense. And I like the idea of using power which nature squanders all over the place. Coal and oil are running out or costing more to get, and the alternative of the peaceful use of atomic power threatens to stock the world with incredibly dangerous atomic waste which will remain dangerous for thousands of years and need the most stringent, expert supervision all that time. Why not wind, water and tide power? The only objection I can see to them, like the objection to the canals, is that they are cheap, efficient, clean and quiet.

Buildings

There is not much I need say about the buildings. They are inadequate but I manage.

Two excellent concrete corn bins in my meal house, holding between them 60–70 sacks of corn, have simplified the problem of grain storage and handling, and reduced the sacks needed at threshing time. Oats and dredge corn or barley for feeding we put straight into cake or meal sacks and keep emptying them into the bins. In this way about 30 to 40 small bags will keep the thresher going, for they keep coming back empty. And small bags make it easy for me to do the work of carrying and tipping them, if necessary single-handed, leaving the rest of the staff free for the threshing. For though I can carry 18 stone and more of corn on my back I am not one of those Samsons who can pick up a $2\frac{1}{4}$ cwt. sack of wheat or 2 cwt. sack of barley off the floor and throw it over a 4 ft.-high wall into the corn bin.

Various boxes and sheds house the pigs and calves and bull or bulls. And two fine new pig-styes, put in by my landlord in what was the old cowshed, accommodate some of the baconers with labour-saving feeding troughs.

The cows are milked on a yard-and-parlour system which is in no way remarkable. The old horse stable is usually full of calves and is often 2 ft. deep in dung. An Essex hammer mill in the meal house grinds the corn for pigs and cattle.

Swill is cooked in an electric boiler which switches itself on and off, holds 18 gallons at a time and saves me hours of time which I used to spend collecting and chopping wood and making up fires under two smaller iron boilers.

The Farmer

The farmer and his family are the most important livestock on the farm. For the motives and methods of its farmers will affect the whole face of a countryside and go a long way to affect the well-being or otherwise of the other people and animals that live there.

I am not farming to dodge death duties or take my losses off supertax, nor am I trying to make the maximum short-term profit at the land's expense. If I were doing any of these things, Pucketty would look very different.

Really I am farming much as our ancestors farmed, for a living and to try to do my duty by the land, as I see it. I call myself an organic farmer, but that is a vague term. For the organic movement contains at least as many varieties of farmer as orthodoxy does. Some organic purists will not consider me an organic farmer at all. But I am certainly not orthodox, and if I am to avoid the stigma of being classed in a minority of one I must beg to be admitted to the fold of the organic movement, even as a black sheep.

So now for a few pages about my ideas.

The discovery of the role played by chemicals in plant nutrition was a great advance in our knowledge about crop husbandry. It led to the modern fertilizer industry and 'orthodox farming'. Up to a point all this is excellent. But it has gone a bit wrong, as I shall show presently.

As a protest against orthodoxy there has grown up the extreme organic school of farming which laughs at the discoveries of the chemists from Liebig onwards and pins its faith to the 'biological fertility' of the soil. The soil is a mass of live organisms, they say. Chemistry has almost nothing to do with it. They do not deny that plants require minerals for their proper growth, but claim that there is plenty of everything in every soil and it only needs to be in biological trim to prove this. Up to a point this, too, is quite right. But these folk who shout 'fragmentation' at the chemists are themselves guilty of fragmentation in concentrating solely on the biological side of fertility.

My position is midway between these two. It is not that I am afraid to come out strongly for one side or the other. But there is truth, partial truth, in both camps and I want to achieve a fusion of the two. Ultimately the only kinds of farming which can succeed and go on succeeding for ever are those which return to the soil the *minerals* removed from it by cropping, and at the same time supply it with humus and keep it biologically healthy and active.

I object to fertilizers, not as a whole class of substances, utterly beyond redemption, but as and when they are the *wrong thing*. Let me give examples. To things like slag and rock phosphate, which are impure, unrefined, largely insoluble and which do not contain, so far as I know, harmful substances, I have no objection.

But sulphate of ammonia, superphosphate and many other

fertilizers are artificially produced, too pure (and therefore lacking in trace elements), highly soluble (and therefore possibly upsetting to the delicate organisms in the soil), and often contain harmful residues. Sulphate of ammonia, for example, when the nitrogen has been used up, leaves the equivalent of sulphuric acid behind. Hence the fact that it makes soil sour unless lime is added to neutralize it. The addition of lime makes gypsum and there is nothing to show that this is necessarily beneficial to the soil.

Superphosphate contains an excess of sulphur over and above what we believe to be needed by crops. And so on. There are probably other objections to these *artificial* fertilizers which we don't yet know about. I like the attempt of the fertilizer industry to put back into the land the minerals removed in crops. But it has failed to do so. It usually puts back a cheap or more highly soluble substitute for what has been removed or for some of it.

The apparent success of orthodox farming has led to an arrogant 'scientific' approach to farming problems generally, especially by the chemists, and the use of artificial fertilizers is, nowadays, the least of the sins of orthodox farming. The use of poison sprays, as insecticides and weed killers, is, in my opinion, far more dangerous. I do not class it as unwise action but as criminal lunacy.

On the other hand I have no sympathy with the out-and-out organic enthusiast who believes that humus is the only need of the soil. I am afraid their school of thought is simply muddle-headed. They argue thus: 'nature produces abundant soil fertility without fertilizers—only with humus. Therefore all we need do is follow her example and all will be well.' So far so good. But they *don't* follow her example. No one in modern England can do so because of an insidious invention called the water closet.

We farmers sell our eggs, meat, milk, corn or vegetables and flowers to the townsman. This is how the townsman survives and how we make a living.

But the townsman puts his dung and urine down a beautifully glazed china drain and thence to the sea as quick as possible, and he throws any waste food (and he loves to waste food) and other organic matter into great 'tips' where it remains useless, and he buries his dead in specially segregated, sterile plots. Now it has not occurred to some organic farmers that this is not how nature goes on.

In nature almost all dung and urine fall back on the land that grew the food which made them. All animals and plants die and their bodies either rot into the same soil, or are eaten by other animals and returned to the soil as excrement. Thus *all minerals eventually go back into the land as well as all humus-forming plant and animal wastes.*

Nature has reserves of minerals in the soil and a pretty large stock of minerals locked up in the bodies of living organisms. These keep changing, passing back to the soil, and then back to plant and animal tissue and so on. There is always a lot at any one time locked up in living tissue. But this is not 'sold off the farm'. These reserves plus those in the soil are very large on some virgin soils, but they are not inexhaustible. There is absolutely no reason why they should be inexhaustible—beyond the fact that it would be very convenient if they were.

Deep-rooting crops supplied with plenty of humus and organic nitrogen can bring up minerals from the subsoil to correct many mineral deficiencies if they are grazed, composted or ploughed in. An oak wood builds up enormous reserves of potash in the leafmould and timber, much of it obtained, perhaps, scores of feet down in the subsoil. If the ashes of the timber are returned to the land or if the timber is just left to rot, as nature would leave it, fertility is stepped up tremendously without any imported manure. But oak takes one hundred and twenty years for a single rotation. And you have only to look at the miserable, stunted oaks on poor heath land to realize that deep roots cannot get minerals where there are none.

As an experiment I have tried to farm some of my heavy clay fields without phosphate. But when they are in grass this takes more muck or more magic that I can supply.

Our predecessors from pre-Roman times onwards had no

artificial manures, but neither had they the modern organic farmer's techniques of applied biology. I may be wrong, but I can't help feeling that had they been so equipped, our land to-day would be much poorer in minerals than it is.

The reader may think me a fool to look ahead for a thousand or even a hundred years. But we must continue to do so. Without hope for the future life becomes impossible. It is only by faith that mankind can hope to survive the twin threats of chemistry and physics.

These views on minerals form the basis of my farming beliefs. I try to return to my soil the minerals which I remove from it, either in organic combination or in crude, unrefined form (e.g. rock phosphate). Although a lot of potash goes off the farm in milk, very little is sold in the bodies of cattle and pigs, and a lot is imported on to the farm in the form of straw, swill and saw-dust, and so I do not have to buy potassic manures. My land, also, is initially rich in potash. If I farmed a poor, potash-deficient sand I might have to choose the most harmless of the artificial sources of potash, if only to get the ball rolling.

I use no artificial nitrogen. A healthy pasture will supply its own needs from the fixation of the legumes in it and the nitrogen returned in rapidly available form in the dung and urine of grazing beasts. Arable land, too, supplies its own from legume crops, the store from when it was in grass, the supplement from dung or compost, and the free-living, nitrogen-fixing bacteria which are found in healthy soils supplied with sufficient humus. I believe that the invitations to top-dress with artificial nitrogen for early bite and so on, advertised so expensively in the farming press, are merely invitations to bad farming. It is bad farming to spend money as a substitute for good management and to the detriment of the land. And it is bad economics to make stuff in a factory which can be made out of doors free of cost.

I further believe that land which is dosed with bag nitrogen comes to expect it and the only hope is to go on giving more and more and more. This is borne out by the amount of fertilizers now needed to produce crops on land which previously needed little or no help from the chemist.

There are no good artificial nitrogenous fertilizers (if you discount soot) and it is these which are responsible for most of the evils that spring from the use of chemical manures. As this element is the most easily replaced of all, even by those who are not organic farmers, there is scope for the saving of millions of pounds on the national scale, and a useful sum for the individual farmer on this alone, even if only the money angle is considered.

I have always felt that nitrochalk was one of the more reasonable fertilizers, that it avoided most of the dangers of sulphate of ammonia and, used in moderation, was probably fairly harmless.

But observations of grazing cattle on land part of which has been treated with nitrochalk have convinced me that it is bad. For it is clear that it makes the herbage in some way distasteful to the cows.

One summer I was grazing a small paddock on fertile soil. I put a fair dose of nitrochalk on about one-third of it. The cows largely avoided this piece of the paddock until nearly all the other two-thirds had been grazed.

In 1954 one of my neighbours sowed nitrochalk on a good grass ley. The fertilizer was applied unevenly in strips of alternating light and heavy doses and it ran out about half-way across the field. The lush, dark green growth where the nitrochalk fell was very noticeable several weeks later when the cows were again turned in.

They grazed all over the field the first day, testing and tasting. After that they grazed almost exclusively on the part of the field which had had no late summer dressing of nitrochalk. Only when this had been grazed to the ground did they come back to the nitrochalk-dressed strips. And even then they grazed most heavily on the strips which had had the lightest dressing of fertilizer, and frequently went back to the undressed side of the field to have another nibble at the bare turf. I have often seen the same man's cows grazing right in the hedge bottoms where the fertilizer drill could not reach, even though there was no shortage of grass in the field.

This aversion to nitrochalk-dressed grass is in marked contrast to their attitude to grass grown with any kind of urine or with chicken dung. In these cases the dark green herbage is greedily eaten and grazed to the roots. (Cattle avoid grass near dung pats, but here it is their sense of smell and not that of taste which is offended.)

When I began to give up artificial fertilizers I never distrusted nitrate of soda as much as, say, sulphate of ammonia. My reasons were all theoretical and apparently excellent. (1) It was a 'natural' product; more or less unrefined and containing trace elements (boron for one). (2) The 'soda' which is left when the nitrate is used up is probably of some use and probably does less harm than the 'sulphate' of sulphate of ammonia. (3) It is highly soluble—but so is the potash in wood ash—a 'fertilizer' whose worth has been proved by centuries of sound tradition. So far so good. But now comes news from America that nitrate of soda is one of the worst fertilizers for causing panning and loss of physical condition in a soil.

To keep a soil healthy it must be supplied with humus. Even the out-and-out chemists admit the need for humus to maintain the physical properties of the soil. I entirely agree. But this is only one property of humus. It is performed by the relatively inert humus which gives the soil its dark brown colour. At least as important is the 'freshly prepared humus' of which Sir Albert Howard was always talking. That is what keeps the soil's living population in trim and on its toes. Soil organisms are always reducing freshly prepared, active humus to the inert dark brown, structureless form of humus, and they must have fresh supplies of organic matter if they are to keep doing their jobs. If they can't keep doing their jobs they just die and the 'biological status' of the soil falls. This is the first stage in loss of fertility. It is easily remedied.

The next stage in loss of fertility is loss of physical health. The crumb structure or friableness and porosity of the soil decline. The last stage is the loss of the soil by erosion. Loss of chemical fertility probably accompanies all three stages, or at least the last two. The killing off of soil organisms—as in fallowing—may raise chemical fertility temporarily.

Our ancestors knew the need for biological fertility even if they never called it that. When they fallowed land (and so killed millions of soil organisms as well as weeds) they gave a coat of dung when the land was clean, and so restored the life of the soil.

They say a little learning is a dangerous thing. But it is not dangerous if it is accompanied by a little humility. The worst thing you can do, when confronted with the mysteries of life or the universe or religion, is to believe you understand anything. Such subjects are far too vast and complex for complete rational understanding. It is far safer to trust your instincts. This does not mean you should cease to seek knowledge or use your intelligence. It means that you should constantly guard against thinking you have found out the last word about anything.

An example is the 'harmlessness' of hormone weed killers. In theory, yes. In practice, no. I am certain they kill bees. But apart from theory and apart from practice, you have only to smell the stuff, you have only to inhale enough and get a real, splitting headache, to be told by your instincts, 'no, not fit for the land'.

The same for A.I. In theory it is the answer to prayer for the small farmer. But no one with good taste would have anything to do with it. Reproduction and sex border on the supreme mystery of life and creation, and to start arrogantly monkeying about with them is bound to bring a terrible retribution sooner or later. Present indications are that it may well be sooner.

So let our motto be: 'understanding with humility'. It is so fatally easy to let your mind persuade you something is reasonable and harmless against your instincts and the verdict of sound traditional practices, and yet your instincts are more reliable—or they will be if you are really a husbandman. If you are not you ought not to be farming.

Ultimately then there are only two justifications for my beliefs and practices. One is that I believe I have good taste and can recognize it in others and follow their good examples. The other is that my farming is more or less traditional. In the last analysis I cannot justify any of my farming on purely theoretical grounds. It works and it satisfies, even though imperfectly, the demands of good taste. And the experience of the past suggests that my methods will go on working.

I make no apology for this. For in anything which really matters taste and tradition are the final and only arbiters. You can settle a problem in mathematics by argument and reasoning and a problem in chemistry by experiment. But mathematics and chemistry, in the ultimate analysis, do not matter two pins to the human spirit.

The faith we hold and the things we believe in (rather than the things we believe) and the way we behave, these things matter enormously and they are entirely a matter of good taste. And about good taste there can be no argument.

So this book is not addressed to the enemy. I do not hope to make organic proselytes of the orthodox heathen. I only hope to show that organic farming as I try to practise it is a workable proposition and that anyone who already has even a sneaking sympathy with it need have no hesitation in going in for it heart and soul.

I said just now that chemistry did not matter to the human spirit. But food does. Not only is food necessary for life and life for the earthly activities of mind and soul, but unhealthy bodies are poor homes for immortal souls and quite unsuitable vehicles for efficient minds.

Food is one of the most important things affecting health. And it can be very easily proved that artificially manured ground produces food quite different in chemical composition from food grown on organically manured land. For colour, taste, texture, keeping qualities and percentage of nitrogen on analysis can all be altered by artificial manuring.

There is a growing body of evidence that such alterations are harmful; that food so grown is less beneficial than food grown organically. This evidence is there for those who care to study it. For the man of taste it will be self-evident that food grown on inert chemicals alone will lack the vigour and palatability of food grown with dung or compost, and I am not going into all this evidence here. What a farmer has to decide is whether he is going to produce the best possible food for the nation or whether he is going to turn out an enormous tonnage of devitalized stuff and sell it as if it were good food in order to make as much money as possible. He has also to consider whether he wants to leave his land healthy and in good heart for his successor or whether he will get rich by making his farm poor and let posterity clear up the mess.

Sir George Stapledon, almost alone among the more orthodox school, has deplored the increasing materialism and mania for money which has now infected the farming community.

Farmers, along with nine-tenths of the rest of us, are now largely preoccupied with money. In part this is the result of criminally and stupidly high taxation. In part it is due to the weakening of the nation's moral fibre and the breakdown of values higher than money values.

I believe that even in competition with orthodox farming organic farming can hold its own. That is to say I believe it can be profitable in the long run, and without recourse to selling pedigree stock at ridiculous prices for their snob value. The ordinary commercial farmer has got to be able to make a living at organic farming if the nation is ever again to have decent food to eat.

My crop yields dropped when I suddenly abandoned artificial manures and it has taken me a long time to build up the land's fertility which was sadly lacking when I took it over just before the war. But recent trends and improving crop yields convince me that the farm is becoming more and more profitable. Paper accounts are not very convincing. But I already bless the day I abandoned even slightly orthodox methods.

I have not made much money in any year so far, and I have sometimes lost it. But my profits have definitely gone back into the farm to a great extent (where they do not wholly show on paper) largely in the form of labour, carting dung, and in rearing healthy heifer calves *on milk* in order to breed my own strain of cow. These are long-term policies which will only prove themselves over a period of years.

But the reason for farming well is that it is right. So long as the right thing to do is still possible it should be done.

CHAPTER II

THE CROPS

Wheat is not my most important crop, but it somehow takes pride of place. Of all the cereals it is the most demanding on soil fertility, and its position as the main cereal of human diet, at least in England, gives it a prestige that none of the other corn crops possess.

The big millers, who dominate the whole field of human food and animal feeding stuffs, declare that most English wheats are 'biscuit wheats', unsuitable for bread-making. They prefer the bone-dry Canadian wheat. It may be grown on exhausted soils, but when it is washed it will take up 5 per cent or more water which need not be dried out again. Hence for every 100 tons of wheat the miller buys he is able to sell 105 tons, 5 being water! No wonder the Canadian wheat is 'more suitable'!

The old stone mills used to make a much better flour with English wheat than with the really hard, flinty wheat of Canadian type.

The baker, as well as the miller, now prefers certain types of 'strong' wheat because they can be made to combine with more air and water and so make more loaves (of less nutritive value) per sack of flour.

Both miller and baker do not hesitate to call in the chemist, who is as willing as ever to produce something which will increase his client's profits at no matter what cost to the nation's health. In any case white flour, even without its poisons, would be very poor stuff.

In these circumstances, and with the government threatening to add chalk to even the wholemeal flour sold in the shops, it is no wonder that more and more people are trying to obtain wheat and make their bread from it under conditions which will guarantee them a wholesome food. It is also significant that these people usually demand wheat grown on organic manures and free from chemicals. For the chemist's interference with 'the staff of life' begins when the wheat crop is sown or even before.

I always try to buy seed wheat (when I buy it at all) undressed with a mercurial seed dressing. Mercury is exceedingly poisonous in some of its compounds, and no one has yet bothered to find out what actually happens to the mercury used in seed dressings. The best guess (it is no more) I have been able to get out of a distinguished chemist is that it accumulates in the soil complex. In any case it must either do that or enter the crop or wash out into the subsoil water which will, in due course, become someone's drinking water. Whichever it does it seems to me that even minute traces may prove deadly. A German scientist claims to have caused cancer of the liver of mice fed minute quantities of a mercury compound of the seed dressing type. If this is confirmed it is very alarming.

I used to retail my wheat to private customers who ground it and made their bread.[1] But the extra work was too much. Hardly any two customers bought the same quantity, and what with confusion about who had and who had not returned their empty bags, and the work of sending out the bills, and the customers who refused to pay even when they got the bills it became a first-class way of wasting time and losing money. I had to give it up. But it is still usually possible to arrange that my wheat goes to one of the millers specializing in whole-meal flour for home bread-making, and not to the white-flour millers.

I can grow crops up to about 32 cwt. to the acre without artificial manures. With them I have grown 38 cwt. to the acre. But I believe it is better to go on building up the farm's organic fertility, and I do not doubt that one day I shall grow wheat crops of 38 cwt. and more to the acre entirely without artificial manures.

Wheat seems to need at least a proportion of clay in the soil

[1] There is a vogue for baking with freshly ground flour. Actually flour which is ten to fourteen days old makes the nicest bread.

if it is to do well. Soils with this essential 'stickiness' can usually be made to grow good wheat crops.

Another major requirement is root room. Shallow soils are not suitable for wheat. And too thick seeding gives disappointing results. The best individual plants in a wheat crop are to be found round the edge of the crop or where the 'plant' is thin, where the surviving individuals have extra root range. This is especially so where the edge of the wheat crop borders a piece of land that is being fallowed, when the cultivations on the neighbouring plot furnish the adjoining wheat plants with extra nitrogen, producing darker foliage and larger heads. I have known severe harrowing thus to produce gigantic heads, bigger than those further out in the field where a heavy dressing of chicken dung had been applied.

I do not always manage to harrow my wheat, but I am sure I ought to do so. Harrowing kills some seedling weeds; though a good crop of wheat will smother almost any weeds on fertile land. Docks will survive (without greatly interfering with the crop) and creeping thistles are a real menace. Apart from them, a good crop will usually win against weed competition, but would be better without it. Harrowing probably helps the crop by letting air in among the surface-feeding roots and promoting nitrogen-fixation in the soil. However, if the tractor leaves a sticky wheel mark the soil is too wet, and if the crop has begun to grow tall it is best left alone. Between these two stages it is not always possible to fit in the harrowing. If the soil is loose and 'hollow', rolling (preferably with light harrows hung on behind the roller to leave a dust mulch) is essential. Under these conditions the tractor wheel is the best roller.

The crop's need for root room and its need for nitrogen seem to be related. I have sometimes found a row of wheat where, due to some defect in drilling, about ten times the usual seed rate was applied to the one row for a few yards. The plants here are sickly and a poor colour. But if a lavish dose of artificial nitrogen is applied to this row they soon acquire the dark green colour of the rest of the field. Unfortunately I have never followed the results through to harvest.

This question of root room may really be, fundamentally, one of water supply. Abundant water and abundant nitrogen have very similar results on a crop. And it is possible that extra root room simply allows the crop to gather extra water.

I have heard it said that a single wheat plant, growing unrestricted and free from competition, will produce 50 miles of roots. Whether that is really true or not, the more root room the crop can be given the better.

Organic matter in the soil increases this root room by increasing the internal 'surface' of the soil. There is rather a tendency among agriculturists to think that heavy soils can get along without much muck, whereas the light soils need abundant humus to offset the effects of drought. But anyone who has seen the effects of abundant organic matter plus the physical properties of clays will never regret anything which increases the organic matter content of his heavy soils.

I have tried various varieties of wheat at different times. The choice has to be restricted to the short, strong-strawed varieties, for on my fairly fertile or very fertile fields any other kinds would be liable to lodge before harvest. At the moment I am growing 'Welcome', a wheat recently introduced by Gartons. Both its parents (Wilma and Jubilee Gem) have done well for me in the past, and Welcome seems to combine the best features of both parents. And it makes lovely bread. This is a big consideration for us, for we use our own home-grown flour for bread, cakes, pastry, etc. We even use it, with the very coarsest bran sieved out of it, as a 'cereal' supplement for babies of 6 to 9 months old. It produces better (and far cheaper) results than the foods sold specially for the purpose.

With spring wheat I have never had much success though I have grown a useful crop of Atle sown at the end of September. In 1955 I grew a good crop of spring-sown Atle.[1] It was a wet spring and a very warm summer and the wheat was on superb land. Even so, winter wheat might well have done better. In 1953 my neighbour grew a really fine crop of Atle. But it is the only really first-class crop of spring wheat I have ever seen in my

[1] 34 cwt. per acre.

5 and 6. Before . . . and after.
(Ten of the piglets belong to this sow.)

7. Lush growth on a herbal ley.

8. Oat stubble on left of picture. On right the same undersown with rape, trefoil and Italian ryegrass, in season of extreme drought.

immediate neighbourhood, and it is well to remember that this was a particularly favourable season when almost all cereals did well everywhere near here.

The great advantages of winter corn are that it ripens while the weather is usually fairly good and the days long. And if a big acreage of winter corn is sown it leaves you free in spring to devote more time to a good acreage of roots and kale. I am also convinced that winter oats, at least, are more nutritious than spring oats. Whether the same applies to other winter crops I cannot say.

We hear a lot nowadays about the ravages of wheat bulb fly, and campaigns are being launched against it. Though I have suffered damage by this pest, the reason is always the same. I have left the land bare of crop in the late summer.

This is just bad farming practice. A cover crop of some sort should be sown—preferably mustard—to be ploughed or disked in later. Then the bulb flies do not lay their eggs in the soil and a fresh dose of humus results from the mustard.

Of course it is not always possible to sow a cover crop. In which case no one can be blamed for the 'bad farming'. But it remains bad—the wrong technique for wheat production. This year, 1953, I sowed wheat on bare fallow where, owing to dry conditions we were unable to sow mustard in time to be any use. A good deal of weed cover had been ploughed down earlier in the summer so the biological status of the soil is likely to be good. But I shall not be surprised if I suffer from wheat bulb fly later on.

Note added in 1954. I did not. The reason probably was that (owing to mole draining the field) I had to fallow it two years running before sowing the wheat. Hence the 'fly' (maggots) probably died out in the spring of 1953 when they found no wheat to eat.

Oats

As with wheat, spring oats seldom do as well with me as winter oats. I have sown my own seed of S.147 winter oats every year now for about six years (seven years counting the first).

They are far from pure now, containing grey and even black specimens. But their stamina and yield seem to be still improving. Though the straw often grows to a great length (1954, 6 ft. and more) it seldom lodges. Altogether I am loth to change to bought seed with more pedigree and less stamina.

My experience with most bought seed of all kinds of cereals is that it is usually short of stamina. It germinates, but that is sometimes as much as many of the seeds can do. When they fail to find their accustomed dose of artificial manure within an inch of their first rootlets, they just fade out. It is not always as bad as that, of course, but the resilience of home-grown, organically manured seed in contrast to bought seed is often very striking.

One disadvantage of organically grown seed is that its flavour is obviously superior, and birds from far and near come to eat it as soon as it is sown and again when it is coming through the ground. This superior tastiness of organically grown crops is a serious matter. It is true that pests tend to avoid them in favour of weakly crops. But the higher animals, rats, rabbits, rooks, pigeons, etc., will go to some trouble to get my crops in preference to my neighbours. This applies to potatoes and fodder beet, kale, cereals, etc. If organic farmers were commoner it would not matter. But at present the vermin do tend to concentrate on us.

Barley

This is not really barley land, and I do not try to grow malting barley. What I do grow is usually a strong-strawed high-yielding type, intended solely for stock feeding. Herta or Abed Kenia will yield well and stand up to a fairly high level of fertility. From 1 to 1½ cwt. of seed sown in a good tilth will usually do well. Sowing both ways almost certainly increases the yield, as it does with all cereals, if time can be found for it. If a ley is mixed with the barley seed and drilled with it, the cross-drilling pays even better. Failing that, harrow across the line of the drilling. On my land the loamy soils at the north end of the farm usually throw up a crop of charlock in the spring, and drilling is best delayed until cultivations have germinated it and killed the first big batch of seedlings. If the creeping thistles are present do *not* use a thistle bar before sowing the barley. It usually gives you three thistles for one.

Dredge Corn

However, my chief spring cereal crop is dredge corn, a mixture of oats and barley, and usually beans or peas, or both, as well. This not only plays for safety, where either oats or barley might do better than the other, but it usually gives a very useful mixture of grains which serves very well for pig or cattle food, and will do poultry as well.

If winter wheat or winter oats are too forward in the spring and look liable to lodge, or if they are just good strong crops and we are short of keep, I graze with the cows, if soil conditions permit. This usually serves to reduce the final height of the crop enough to avoid lodging.

Beans

Beans are regarded as a chancy crop. Winter beans do pretty well for me if the rooks leave them alone. This is partly a question of season. In cold, dry autumns they are very bad. But in wet, mild seasons they are usually happier on the grassland and do not trouble arable crops so much. But it is seldom wise to sow beans on an outlying field, unless they can be sown really early (September for preference). Even if rooks leave them alone the crows will very likely break off the young shoots just for mischief.

Turnips

I do not grow white turnips as a regular crop—only as an occasional catch crop. But I always like to have a few rows of swedes. They seem to do better in the fields than in the garden, and we eat a great deal of swedes both cooked and raw during the winter. Any that are surplus to our needs in the house are usually fed to the calves.

Potatoes

Here again I grow potatoes in the field chiefly in order to ensure an adequate supply for my family.

I used to grow them on a larger scale ($\frac{1}{2}$ to 1 acre) as a contribution to the nation's iron rations in war time. Now I leave that once more to the professional potato-growers, though, in my opinion, their products are, often enough, scarcely fit to give to the pigs. They are grown, often with no attempt to supply humus, on soils hopelessly over-dosed with artificial, and the resulting crop, though it may be heavy, has nothing else to commend it whatever. They taste horrible and turn black when cooked, and their skins are uneatable.

I have one or two customers, including the agent for a big firm of seedsmen, who must know scores of potato-growers, who rely on me for eatable, tasty potatoes for their families.

The chief problem is seed. Where to find healthy seed. The rubbish sold as Scotch seed (there are, of course, exceptional, good samples) may carry a certificate to show that it is free from more than a small amount of virus disease, but it succumbs to blight very easily unless repeatedly sprayed with poisons, and dry rot is usually rife in every bag.

But in 1953 I obtained some seed of Golden Wonder compost-grown in Scotland, by my friend, Mr. Donald MacLean. The vigour of these potatoes grown alongside Home Guard and Gladstone seed obtained commercially was very striking. And the resistance to blight was most encouraging.

Potatoes are one of the few farm crops to which I almost invariably apply compost. Now that I no longer make a lot of compost I keep some of what I do make for this vital crop, and so obtain healthy, well-flavoured tubers and a good yield.

The Golden Wonders I referred to above yielded at the rate of 13 tons per acre. And I have in the past had Gladstones which yielded, at a rough guess, 17 tons or more per acre. At this rate it is obvious that I do not have to put in many rows to feed my family and two or three others. In unfavourable seasons I get lower yields.

The potato glories in loose soil. It flourishes in it and it grows luxuriantly in pure compost. I know that some people grow potatoes without digging. I believe it can be made to work. I always try to give mine loose, loamy soil and old rotten dung or compost to make it even more friable.

I once grew some potatoes by accident in a cooling-off compost heap. They 'ran' like squitch grass with no attempt to make tubers, only more tops and then more runners. Eventually they did make tubers and a very heavy yield per plant would have resulted if the autumn frosts had held off a little longer.

Wood ashes sprinkled up the drills (and/or ground chalk) at planting time help the yield and certainly do not damage the quality.

Kale and Rape

Rape I grow as a catch crop or in rare emergencies when a crop of kale has failed, which it almost never does.

The chief thing about kale is to dodge the turnip 'fly' or flea-beetle and to give the crop plenty of dung. In 1953, a highly favourable year for root and green crops, I have seen several crops of kale where no dung was given and they have not been good. One or two were quite pathetic, hardly worth running sheep or cattle over.

If kale follows newly ploughed ley, dung may not be necessary, and of course if compost can be given, dung will not be needed, But it is a crop that can scarcely have too much humus.

This business of organic manure and not artificials seems to be the essential thing in beating the 'fly'. And the more intimately the manure is mixed into the surface soil the better. This is confirmed by George Henderson's method of establishing kale, with a very rich dose of peat-litter, chicken droppings and pigs' urine applied as a top dressing and harrowed in before drilling the kale.

I don't want to boast, for I realize that I am extremely lucky with kale. But when that luck seems to turn up, year after year, I begin to wonder if there is something more than luck involved. I sometimes get a bad patch in a piece of kale, but hardly ever a real failure. In 1954 pigeons ate the kale for months after it was *sown* until laid corn arrived at the end of July.

My success is partly due to the fact that it almost always rains when I sow kale. This is luck. But it happens so regularly that if I sow when my neighbours are haymaking I feel quite guilty about it!

But rain is not the entire secret of my success with kale as the following experience with rape will show. One summer I bastard-fallowed two bits of ground. They were in separate fields. One piece was mucked with all sorts of odds and ends of stuff—chicken dung, genuine compost and old farmyard manure. But it all got a good dressing which was well worked into the soil. The other field received no manure.

When wet weather came at the end of July I drilled rape on both pieces. Rain was incessant. Scarcely a day went by without at least one shower. Conditions seemed perfect. Both pieces of rape established themselves well, so that they were beginning to 'change leaf' and could be seen right up the field in the drills.

The dunged piece never looked back. A few flea-beetle punctures appeared in the young leaves, but it was nothing to worry about.

But the other piece was attacked by 'fly' in a really astonishing manner. In about three days they wiped out almost every single plant on about 8 acres of rape. They nearly destroyed some nearby plants of kale which had grown up from kale stubble in a fallow after kale. These were nearly a foot high, but were almost annihilated.

That the soil was chemically fertile was proved by the way Italian ryegrass established itself and grew following the rape failure. But I suspect that this is not enough. The fallowing had impared the biological fertility of the soil, and it was not restored in the case of the field not dunged.

Of course the crop that failed *may* have been attacked simply because there was a high and very local flea-beetle population which may not have been the case with the other field. But taking this case in the light of much other experience and observation of 'fly' attack I am inclined to think that the biological status of the soil is the supreme factor. This agrees with the fact that farmers who dung their kale, but add a lot of artificial, often have to resort to insecticides to establish a plant (they do not always succeed even then) and this *at the same time* that my kale is growing away from the 'fly'.

If I am troubled by fly and hot, dry weather at the same time I like to dust the kale seedlings with a little basic slag or ground chalk. This seems to discourage the fly. I never use real insecticides, certainly not D.D.T.

I heard an amusing story about a farmer who is also a partner in a firm of seedsmen and agricultural merchants.

Flea-beetle was ravaging his 12 acres of kale, so he rushed off with car and trailer to the shop and demanded the appropriate flea-beetle dust. They offered to send a man along with him to the warehouse. But he was a partner in the firm. He knew all about it. He could help himself. They told him where it was and he rushed off and seized the appropriate number of bags, flung the evil-smelling stuff into his trailer and dashed for home.

It was instantly given to his tractor driver who raced with the manure distributor to cover the crop before dark.

Only later, when the entire crop vanished without trace, did someone look carefully at the bags that had contained it. They had used the hormone weed-killer for charlock and other brassica weeds!

I usually grow thousand-head kale as the cows seem to prefer it to the marrow-stemmed variety. I certainly prefer it. It produces more leaf and less stalk and it retains its leaf better in frosty weather and is much hardier than marrow-stem.

Marrow-stem is usually all right if it is to be used in the autumn, though the unusually savage November in 1952 made a lot of marrow-stem kale look silly. Where the crop is for use mainly after Christmas thousand-head is much preferable. The short severe frost in 1954 (January and February) destroyed thousand-head kale, but only where pigeons also attacked it, rather as grass will go black in a frost only where it is trodden on.

I usually grow several acres of kale and fold the cows over it with an electric wire. We seldom have time to single the plants. I am sure that, if we did so, we should get heavier yields and more leaves, more thousands of heads in fact. But as it is, yields often come out about 30 tons to the acre, and that is nothing really to grumble at.

I confess that I have not sampled my kale crops with the same zealous accuracy and impartiality which I show with fodder beet. But on a cold autumn day there is little pleasure in plunging through kale 5 ft. or more in height every leaf of which gives you a miniature shower-bath.

This height sounds wonderful. But actually I would rather have a crop 3 to 4 ft. high and thickly massed with heads—in other words a well thinned crop. A friend of mine grew a crop in 1953 even taller than mine. But he agrees with me that the extra height is all stalk where the crop is unthinned. This is no advantage and may be a nuisance for electric fencing, and also cause the crop to lodge.

Fodder Beet

My only other crops apart from leys and catch crops are fodder beet and mangolds. I am not sure yet if it pays to grow mangolds as well. 1953 was the first year in which I have grown fodder beet, and I will not pose as an expert.

I had had it so drummed into me that fodder beet was a greedy feeder and needed a lot of artificial that I put up a home-made complete artificial that really *was* complete for two plots. The mixture contained about 1 cwt. per acre of real N.P.K. artificial and the rest was slag, ground chalk, wood, coal and peat ash, a very little fish manure and some bone ash and steamed bone flour and granite dust. That I reckoned should be a really complete 'artificial', with all trace elements known to science. I gave 5 cwt. per acre (approximately) to two pieces of beet and none at all to the third. All three had dung ploughed down under them and the third (no artificials) plot had a dressing of compost harrowed into the soil after ploughing in the dung.

This last plot out-yielded the other two, although sown over a fortnight later. This extra yield was probably due to closer spacing of the roots and not to the absence of the mineral mixture. But it has proved to me that fodder beet is no exception, and like other crops, can be grown very well on organic manures alone.

The actual yields were as follows:

Pajberg Rex X (a gappy plant) just over 20 tons per acre.

Yellow Daeno X (a good plant) about 30 tons per acre.

Yellow Daeno with closer spacing and dung and compost only (an excellent plant) 33 tons per acre, with about 21 tons of tops per acre.

Actually this last piece had a fair proportion of Pajberg Rex X roots among the Daeno. Had it been a pure stand of Daeno the yield would undoubtedly have been even heavier, though there might have been slightly less tops.

The Yellow Daeno seems to be excellent (and is recommended as being suitable) for feeding to either cows or pigs, and as it lifts very easily and yields better than the white varieties, and produces a better shaped root with less tare on it, I think I may well restrict my choice of fodder beet another year to this variety.

With 50 tons or more of roots and tops per acre, all good food, it is difficult to see what crop can beat it on a yield-per-acre basis. Yellow Daeno X is said to produce more dry matter per acre than any other variety of fodder beet. And since it combines this high yield with easy lifting there is a lot to be said for it. The higher percentage of dry matter in Pajberg Rex X is partly due to much higher fibre content, as anyone can tell by splitting the roots.

This year, 1954, I am trying mangolds versus fodder beet to see which will yield most food per acre.

Kale is the only rival to fodder beet in my view. Its yield of food per acre may be lower but it produces more protein and green leaf per acre, and is less trouble to weed.

For pigs I suppose fodder beet, on the whole, is better. But for cattle I think a bit of both is ideal, the fodder beet as a supplement to the kale and a substitute for it when the kale is frozen and cannot be fed.

One crop I never grow is fodder maize. I cannot see the point of it unless it is to be ensiled or fed to stock indoors. It takes about as much manuring and attention as kale and is not ready for cutting till late summer or early autumn when there is almost always a flush of grass. It is low in protein and therefore not a great help with dairy cows and as it is frost-tender it is of no use whatsoever as a winter feed unless it can be chopped and ensiled.

Year after year I see fine crops of maize killed by frost before they can be used. It seems a great waste of effort and good land.

Silage Crops

I used to make silage with oats and vetches, preferably winter sown. This is an excellent plan in many ways. It is a smothering crop which helps to clean the land and leaves it free in the height of summer for further cleaning with cultivations or, if clean enough, for a crop of kale or rape or white turnips or a ley. I would never sow the ley under the oats and vetches unless I was sure of cutting them for silage before they exert their full smothering effect.

One snag is that there is often a drought which prevents the sowing of kale in time to do any good, so that the theory of two winter-feed crops in one summer off the same piece of land does not always work out.

Another snag, and a worse one, is that on a small farm with a limited staff silage just cannot be made in any quantity at a time when the fodder beet and main crop kale need hoeing and when there may also be good haymaking weather.

I am a great believer in the beneficial effects of silage on milk-yield and health of cattle. But I also believe in hay. (People who do not believe in hay usually turn out to be those who cannot make it in catchy weather.) And it is no good trying to make hay and silage and weed the fodder beet all at the same time. So silage is rather a thing of the past with me.

Rotations

I don't believe in strict rotations. That is not to say that I believe in taking wheat after wheat or that I do not recognize the wisdom of alternating cleaning and fouling crops and so on. But I do not stick strictly to a four- or five-course rotation or anything like that.

On a small farm, even more than on a large one, circumstances vary from year to year. The weather, the loss of a man, a breakdown to some vital machine, not to mention changes in national policies for farming—any of these can upset a strict rotation hopelessly. If I cannot get my wheat sown at the right time it is no good saying, 'very well then it must be spring wheat'. I would change it to spring oats or dredge corn.

Again it is no use saying this field shall be kale in 1958, because a late spring which holds up other work followed by an early drought may make it quite impossible to grow kale there that year.

Disasters are not the only things which may upset a rotation. A bit of hot dry weather in April may enable you to clean a dirty field well enough in a few weeks to be able to sow a crop where you had expected to have to use a long bastard fallow and then sow mustard. So you now sow late barley or Italian ryegrass or, if you can dung the ground, perhaps kale or rape, or you may sow a ley there in May if you get showery weather.

Then again I often get a self-sown ley. I do not mean a tumbled-down long ley. But after seeded red clover, followed by wheat or spring corn, it often happens that the shed seed produces another excellent plant of clover under the corn. When this happens I welcome it instead of solemnly ploughing it in because that piece has *got* to be roots next year.

So long as a farmer has a conscience and a real care for his land an arbitary rotation is not necessary to maintain fertility. The previous history of a field must guide him very largely in what he next grows there.

If the field has grown more than its share of corn crops then it is probably due for a bastard fallow and a three to four years' ley. But if it is a field which has been in grass for years and has only recently been ploughed, and if it is also so near the farm buildings that it gets more than its share of dung, then it can probably be made to grow two, three or perhaps even four corn crops running.

What I believe is far more important than strict rotations is the overall picture of the farm's fertility. Is the land on the farm as a whole getting richer or poorer? I shall have more to say on this aspect of fertility later.

CHAPTER III

LEYS

'Of all cultivating agencies, then, roots stand by far at the head, and it is by applying this principle to our arable lands that we shall at once manure, aerate, and cultivate them in the cheapest manner . . .'

ROBERT H. ELLIOT

The quotation at the head of this chapter shows that Elliot was miles ahead of his time in his conception of ley farming. For in this passage quoted he is referring not to roots in the sense of turnips and mangolds, but to the roots of all farm crops, and in particular of course, ley plants. He was making a plea for the ley on arable land as the cheapest and most effective means of improving soil texture.

This aspect of leys is taken almost for granted nowadays, but I doubt if the orthodox type of ley makes the fullest use of this 'cultivating' and manuring influence of plant roots. The orthodox ley usually consists of a very simple mixture of grasses and clovers, often of only one grass and one clover. But to make the best use of the long ley it is necessary to sow deep-rooting and tap-rooting plants, so that the greatest possible depth of soil is permeated by their roots. And it is sensible to sow a variety of herbs to ensure the health of the grazing animals and the palatability of the herbage.

These herbs probably benefit the soil, too, toning up the soil organisms and making a better humus when ploughed in. Bacteriological work by the Soil Association at Haughley suggests that phosphate-dissolving bacteria thrive best in compost made from a big variety of different wastes. Similarly the humus made from a mixture of herbs and grasses may well be much more beneficial than that made from one grass and one clover. For this reason it has always been my practice to sow the herbs in the mixture all over the field and not in that quaint thing called a herb strip. Anyone who has watched cattle grazing will realize that from their point of view, too, it is better to have the herbs scattered all over the field. Why should a cow have to walk to the herb strip to get what she wants instead of selecting it as she grazes anywhere in the field? The latter way must make it easier for the cow to get a real mixture of herbs throughout her rumen's contents.

I also like, in the long ley, to sow more than one type of legume. Doubtless wild white clover (or one of its close relations such as S.100 or New Zealand white clover) will provide the bulk of the nitrogen-fixing root nodules on which the fertility of the ley will ultimately depend. But that is no reason for excluding other plants, especially in the first year or two.

Only in the matter of the actual grass am I inclined to restrict my choice to one or two species. It makes for easier and more efficient management of the ley if only cocksfoot or timothy and meadow fescue are sown, and not a mixture of several species. Though even here I fancy the cattle would prefer a mixture such as they get in an old pasture.

Here, then, is a typical ley mixture as sown at Pucketty, adapted from the Clifton Park System of Farming:

14 lb. Cocksfoot
1 lb. Ribgrass
3 lb. Chicory
5 lb. Burnett
4 oz. Yarrow
2 lb. S.100 or New Zealand white clover

If the land does not grow wild white clover naturally, it would be as well to include about ¼ lb. of this variety to take over from the less persistent S.100 or New Zealand strains as they die out. White clover can often be introduced to a field from which it is absent by grazing that field alternately with one in which white clover has been allowed to go to seed. Thus the cattle eat white clover seed, say, during the day and deposit dung containing the seed at night in the other field. Harrowing will spread the seed still more and it is sure to thrive as it finds itself in a bed of humus made from the cow dung while it is germinating.

Timothy can replace the cocksfoot and so can meadow fescue or a mixture of timothy and meadow fescue can be used.

Perennial ryegrass is a species I am not much in love with. For an early start it seems to have little if any advantage over cocksfoot or timothy and it is left far behind by its relatives, the H.I. and Italian varieties of temporary ryegrass. And this may be heresy and just due to my bad management, but I find the temporary strains far leafier than any variety of perennial ryegrass.

I think the mixture above would be improved by the addition of a few pounds of late flowering red clover or the Aberystwyth S.123 which is fairly persistent. It will last for two or three years. A little trefoil, too, would not come amiss.

Lucerne is not much use on my land in general purpose leys. It needs to be sown alone or with one, severely limited, companion grass and a little (a very little) white clover. Sanfoin I would sow, also kidney vetch, if I had light, stony or chalky land.

Sanfoin and kidney vetch each have some reputation as herbs, and being legumes and tap-rooters they are bound to increase the general level of fertility in the ley.

Meadow fescue is at least as good a drought-resister as cocksfoot on my farm, as I found out in the phenomenal drought of 1949.

Ribgrass is said to be rich in lime and phosphate, and yarrow is supposed to increase the butterfat in the milk. Whether it does or not I can't say, as I have never grown it alone.

Chicory should be avoided if the field is to be mown for hay. The tall thick stem is a nuisance. But dandelions make good hay as well as grazing and are probably much more medicinal than chicory though they don't produce so much bulk. The dandelion, however, is a true pasture plant and will outlast chicory, at least on my heavy soils.

To save the seed of dandelion, which is rather hard to buy, do as follows. Pick the heads just before they open out into 'clocks'. That is to say when the seeds are brown and the 'pappus' shows as a white nose at the end of the head. With a little practice you can usually spot the riper heads. Then pinch off the fluffy end of the head, holding the seed end tightly with the other hand as you do so. In this way almost all the seeds can be left undisturbed on their 'pin-cushion' and the tiresome fluff can be got rid of before it takes control.

Leave the 'pincushions' of seed to dry in a sunny window and when dry rub off the seeds. This is laborious but worth while if no other source of dandelion seed is available. Of course it ought to be possible to harvest them from a sort of outdoor vacuum cleaner when they are in the fully ripe state. But few, if any, seedsmen yet realize the tremendous value of this plant. Probably no other herb has such a well-deserved reputation for putting a bloom on a horse's coat. And probably the dandelion as much as anything is responsible for the healthy dappling that appears in my red cows' coats in summer time.

Burnet resists drought almost as well as chicory. It has a tap root, about as thick as a pencil, which goes straight down into the most uncharitable clay. It is fairly wintergreen, and I believe it has an excellent effect on the health of stock. Near my farm is a field, now ploughed up, which was mown for hay every year and never manured in any way. Year after year it produced a fair crop of hay. The herbage contained a very high proportion of burnet. This indigenous burnet is quite a different affair from the commercial burnet which is produced as an impurity in sainfoin seed. It is a smaller-growing plant and probably less productive but more persistent.

Yarrow resists drought well. But so far as I can discover it does not form deep tap roots on my land. It is a constituent of many old pastures and once established is rather hard to get rid of by ploughing up. The seed is very small so that a little goes a long way.

9. Mustard growing on dunged fallow. Boy's height about 4 ft. 3 in.

10. The same crop being chain-harrowed and ploughed in October.

11. Flourishing fodder beet.

12. Welcome wheat, 1954.

S.100 white clover resists drought surprisingly well though not like chicory or burnet.

Here is another ley mixture of mine as actually sown:

4 lb. S.143 Cocksfoot
2 lb. S.26 ,,
4 lb. S.37 ,,
4 lb. Sweet Clover (Mellilotus alba)
2 oz. Yarrow
5 lb. Burnet
3 lb. Chicory
1 lb. Ribgrass
2 lb. S.100 White Clover
2 lb. S.123 Red Clover

I am inclined to think now that more chicory should be sown. It has a fairly large seed, and I doubt if 4 lb. or 5 lb. would be too much. The seed is not so expensive now as it used to be when interest in Elliot's mixtures first revived during the Nazi war.

Sweet clover is another tap-rooting, drought-resisting legume. It is a biennial and only lasts for one harvest year unless it sets seed. But it does a lot of work in that time, and the other plants will fill in the gaps in the next year. Cattle and horses will graze it readily when it is young. When it grows taller it is apt to be neglected—probably it grows bitter in taste.

In hay and silage it often causes trouble. For if it grows even an invisible amount of mould it is said to cause failure of the blood to clot in cattle which eat it. If ever allowed to seed it becomes a persistent weed, and if it cannot be kept well-grazed or topped regularly with the mower it is probably better omitted.

I think the real use for this crop is as green manure. One or two big crops of it could be grown and mown before they set seed. Then the tall, stemmy material could be ploughed or disked in and would make a lot of humus. And the leguminous tap roots would be almost as good for the soil as well-established lucerne. Most of my leys are, in theory, 4-year leys, but they can be broken up after 3, 5 or 6 years if it suits you to do so.

I have described in the chapter on catch crops the Italian ryegrass-trefoil mixture I sow as such. It often gets left as a 1-year ley, being grazed till the end of April, mown to yield about 30–40 cwt. of hay in June or July and grazed thereafter until it is possible to plough it up.

Besides this I usually sow one field each year to a ley of broad red clover. About 12–14 lb. of seed is all that is necessary where conditions of tilth and fertility are good. This is sown under a cereal nurse crop. The clover seed and corn can be mixed and drilled together to save time so long as the drill is not set too deep. In this case it is worth while to drill half the seed one way and half at right angles to it. A better plant of clover will result and a better yield of corn too, in all probability, as the young plants each get a better share of soil-water and plant foods.

If the red clover ley can be grazed in the spring I like to sow about 10 lb. of red clover and 6–8 lb. Italian ryegrass. The ryegrass gives a useful early bite before the clover begins to move. Then the two can be left to grow together for hay. If the ryegrass cannot be grazed, for any reason, it is best omitted. Starting growth early it will smother the still dormant red clover and the quality of the hay will suffer. So will the fertility-building of the ley, for less clover top will mean less clover root and that means less nitrogen. The wood-pigeon, rapidly becoming Berkshire's number one pest, will graze the red clover in spring and neglect the ryegrass. So it is even more necessary to correct this with stock that will eat off the grass.

Italian ryegrass does not make good hay. It bleaches before the clover is fit, and it is apt to be stalky, though not so stalky as perennial ryegrass. If an early bite is not needed and if you want the maximum fertility effect from the ley, sow pure red clover. The hay or silage will be rich in protein (if the leaf is not lost in haymaking) and the soil will benefit from the subsoiling of the clover's tap roots and the nitrogen-fixing of its root nodules. A good crop of wheat should follow.

If red clover has been taken too often on the same field, clover sickness may set in and spoil the plant between one summer and the next. If there is any danger of this, sow about 4–5 lb. of trefoil seed per acre or a little Alsike as an insurance against a failure of the red clover. Trefoil is one of the most fool-proof crops to sow. Its rather big seeds all seem to take and the result is that even 4 lb. per acre will give a good proportion of clover in the hay.

So far I have not had much use for 'permanent' or very long leys. But I think for them rather more grass species should be included in the mixture. On my naturally grassy farm more grasses always appear in the long ley before its four years are out than were sown there originally. And if a ley is to stay down for many years there is a good deal to be said for sowing the best possible pedigree strains of several species at the outset rather than leave a vacancy in the mixture for volunteer grasses of poor, commercial strains.

If I am sowing a ley under a nurse crop I broadcast the seed on the surface on a fairly fine tilth and either harrow and roll or just roll. Or I roll first with a ring roller, sow the seed and roll the other way with the ring roller to bury the seed. March and April or July and August are probably the safest months for direct sowing of leys as drought is the worst enemy of young grass and clover seedlings.

To broadcast the seed I either sow by hand, a job I do myself as it is rather skilled, or mix the seed with sawdust and sow it with the slag drill. This method allows slag or rock phosphate to be mixed with the sawdust too, if the field is phosphate-deficient. Slag sown with slightly damp sawdust is much pleasanter than slag sown alone. But on land which is already in a high state of fertility this will not be necessary to ensure a take of the seeds. On some of my poorer clay fields phosphate helps to establish the ley and a dose of it a year or two later will give the ley another lease of life.

If dung or compost can be applied during the life of the ley so much the better. Failing that, red clover straw after the seed is threshed, or bean straw scattered over the field will help to add to the humus the ley is building up and will also supply some plant foods in organic combination. And clover and bean straw, being more nitrogenous than cereal straws, will rot easily and quickly without demanding so much soil nitrogen to do so.

The scientists say that a good ley will add 1 ton of *dry* organic matter to the soil per acre, per year. This presumably applies to an orthodox ley. I imagine a ley with plenty of deep-rooting plants would do better, since a greater depth of soil is permeated by plant roots, yet the surface soil is just as full as with a shallow-rooted ley.

If green crops in general contain, let us say, about 10 per cent dry matter and 90 per cent water, then a 4-year ley will put into the soil in its time as much dry plant material as a 40-ton green crop of kale or mustard. This is a better way to assess the figures quoted for leys than to compare them with dung or compost, since they are at least partly rotted and have lost bulk in the process. But 40 tons of kale or mustard ploughed in would be a formidable addition of organic matter, and it is no wonder that leys improve the soil, especially as their roots also break up the soil in a purely mechanical way and force a crumb-structure upon it.

I suppose actually a 40-ton crop of kale would have nearly 40 tons of roots and roothairs below ground, perhaps more, and so the ley should really be compared with a 20-ton crop.

Conversely this means that a 20-ton-per-acre crop of mustard, grown in a few weeks, supplies in one dose as much humus-forming material as a ley will supply in four years. Actually unless the mustard is allowed to grow stemmy this is probably not true, since young mustard is nearly all water, and ley roots must be much more fibrous. A lot more humus can be made from a 1-year ley by allowing the second crop or aftermath to grow and then ploughing it in instead of cutting or grazing it.

The after management of the ley is important. If artificial nitrogenous top-dressings are to be avoided a proper balance must be kept between the grasses and clovers. Excessive grazing of the sward will dwarf the chicory and grasses and encourage the white clover. This will make a lot of nitrogen but, unless the ley is rested to allow the grasses, chicory, etc., to grow, little use will be made of this nitrogen.

On the other hand insufficient grazing (or insufficient phosphate in the soil) will mean stunted clovers which means lack of nitrogen and starving grasses. So periods of grazing to help the clovers must alternate with periods of rest to let the taller plants grow. This is one way in which red clover, sweet clover and trefoil and other tall-growing legumes come in. For they, too, benefit from periods of rest, and they will go on fixing nitrogen even if some of the white clover gets a bit stifled by excessive growth. Whenever a clover plant is defoliated by grazing it is said to release nitrogen from its root nodules into the soil. This then becomes available for the grasses and other non-leguminous plants in the ley, and if rapid grazing on the graze-and-rest principle is the rule, the rest period (closely following rapid grazing bare of the pasture) will enable the grasses to use this nitrogen for a quick burst of fresh growth.

If at any time during the summer the pastures grow away from the cattle, so that they neglect patches of grass and allow them to grow tall, the mowing machine should be used. In this way it is often possible to save an extra ton or two of quite useful hay from the pastures. If electric fencing and strip grazing are practised this is not likely to happen, but all is not lost if it does happen in a fairly large field. After mowing, the grass grows short and palatable and will be grazed properly.

Incidentally, in connection with the figures I have quoted for dry matter per acre made by leys, a lot of this must inevitably be root material. And heavy, continuous grazing of any crop will restrict root growth. No crop can grow without ample leaf above ground to make its food, and it can't grow roots without leaves any more that it can grow leaves without roots. Each helps the other and periods of rest are needed, even by pasture grasses, if maximum root and leaf growth is to be possible. Only the very flat, low-to-the-ground white clovers are encouraged by continuous close grazing, since this discourages their rivals, the grasses and tall plants, more than themselves.

Leys should be sown in very clean ground if possible. Troublesome weeds can be repeatedly cut with the mower if the cattle won't eat them—weeds, that is, which appear after the ley is sown. Creeping thistle is best mown early in July when the flowers open, cutting as low as possible to the ground, unless repeated cutting can be managed. This one cutting if rightly timed and well done does maximum damage to the plants. Since a huge patch of creeping thistle may be all male or all female, having sprung from one original plant, this kind of thistle often fails to set seed. It can and does seed, however, upon occasion.

The Scotch or 'Jack' thistle should be cut just before the flowers open. Left longer it will seed after it is cut unless very dry weather follows at once. Cut too early it will come again from the root and manage to set seed before severe frost sets in. Once it has set seed and beaten you, it is best left to the goldfinches who will get a big percentage of the seeds if the plant is left standing.

If only a few Jack thistles have to be got rid of, put a teaspoonful of nitro-chalk into the heart of each one in dry weather in April before they begin to grow tall. The thistle will vanish without trace. This is the best use I know for nitro-chalk.

The technique of sowing a ley direct, without a nurse crop, is fairly well known, including the need to graze as soon as possible, say in six to eight weeks from sowing. But a point worth making concerns the use of oats as a means of increasing the first and second grazings. Sown with the ley seeds the oats give a very useful grazing long before much could be expected of the ley, and a second and even a third grazing, though diminishing in size, can be expected from the oats before the ley takes over completely. But in dry weather, when the rooks are very hungry, it can be disastrous for the ley to sow oats with it. The rooks will dig for the oats and in doing so will uproot the young grass and clover seedlings and kill them, and a badly gappy plant may result.

I mentioned earlier the importance of preserving the nutritious leaf of red clover. And the same applies to all hay plants, for the leaf is the most valuable part.

The usual practice at haymaking time in my district is as follows. The grass or clover is cut, a large area being knocked down at one time. Two or three days later this is turned with a royal disregard for the weather. After it has got wet the crop is turned again. Wetted once more it is turned once more and so on until sunshine makes haymaking foolproof. By this time meadow hay is brown and has little smell and presumably not much taste either. Seeds hay, however, consists only of yellow or brown stalks of clover (or lucerne) from which every shred of leaf has long been banished. It commands a higher price than meadow hay. Heaven knows what it is used for.

A better plan is the one I have worked out to try and cope with our climate. I cut no more grass in one day than I think we can save in one day. This is allowed to lie until it is partly dried and the weather looks as if it will hold for the rest of that day. I never trust tomorrow's weather no matter how settled it may look to-day. Above all I never listen to the Air Ministry's pathetic forecasts, not during haymaking time.

When I judge we have as good a day as can be expected we turn the crop as soon as the dew is off it. Then, as soon as it is fit to cock, the tractor haysweep goes in and sweeps up rows of heaps of hay. All available hands then go out and cock the hay, working along the lines of swept hay. Small cocks may be needed at first if the hay is still very sappy. But often, by the end of the day, big ones can be made. They are generally better in every way.

If the weather is windy the hay can be swept to the most sheltered part of the field and the cocks built near the hedge. After they have settled they can be swept into the open to dry faster (so long as the aftermath grass is dry) and can be mowed as often as needful to save the crop beneath them from being killed. They remain weather-proof. Finally they can be swept to the rick or to a baler or big ones can be dragged home with a rope or chain round them and a tractor or two horses on the job.

The plan I follow then is quite simple: leave the hay undisturbed in the swath (when the rain will do it very little damage for a few days) or safe in cocks. In this way it is possible to beat the weather almost every time. Between 1939 when I lost a whole field of hay (following the methods in use generally in this district) and 1954 (when I lost another 7 acres of hay) I never lost more than a few wagon loads of hay, and often saved superb stuff when other farmers with similar crops, and working on the same days, produced rain-sodden rubbish. But 1954 beat me with that one field. I still don't know how it could have been saved. The weather was quite impossible for three weeks or a month. But this is very rare, and though three seasons out of four are catchy and difficult, they are not so difficult that my way of haymaking won't succeed.

Tripods can be used and enable slightly greener material to be used than can be put into all but the smallest cocks. But I don't like tripods if they can be avoided. They are a nuisance to carry about the hayfield and they make the resulting cocks much less easily mobile.

But to save clover seed in a wet September or October tripods are essential. They can be home-made from hedge-row timber.

Quite a good plan with a thick, green hay crop is to build cocks over very small tripods. They need only be about 2 ft. high when erected. They serve to keep the main part of the bottom of the cock off the ground and this means quicker drying and less waste if the hay has to stay a long time in the field. So far I have only used a very few of these as an experiment. But it worked excellently.

The pick-up baler may be useful in some cases. But it is not always possible to dry hay enough for the baler in one day and if more than one day is involved the safety of my system is greatly reduced. All the same, pick-up baling saves labour and a lot of good hay is made that way now. In 1955, with its uninterrupted, good haymaking weather I got a contractor to bale *all* my hay with a pick-up and we never had the bother of cocking any. But it was a quite exceptional season.

I like to chain-harrow in spring to spread dung pats and mole-hills. Later in the year I fancy chain-harrowing just makes it impossible for cattle to avoid dung-soiled grass with consequently increased danger from parasitic worms. Chain-harrowing in autumn is particularly undesirable in fields to be stocked during the winter. In winter cattle graze very close and will eat even unpalatable or slightly dung-tainted grass and if it has been soiled by chain-harrowing worm infestation is likely to be higher than otherwise.

CHAPTER IV

THE CATTLE

Mixed stocking is certainly best from the point of view of controlling the stomach parasites of grazing beasts. It is probably also best from the manurial point of view. Though I have a fair variety of stock, the pigs and cattle supply the bulk of the animal manure and the cattle supply more than the pigs. In summer time I may have half a million or more honey bees adding their quota of animal excrement to the soil, especially in my orchard, but I suppose one cow has them all beat for volume.

I used to serve my cows to calve in the autumn so that most of my milk was sold during the winter when prices were high. But winter prices have fallen lately and I doubt if winter milk, which is so much more expensive to produce, pays any better than summer milk. So I now have cows calving in spring as well as in late summer and autumn. One advantage of this is that some of the calves can be reared while milk is cheap. With abundant good grass it is possible to get 4 to 6 gallons and maintenance off the grass for part of April, May, June, July and part of August. The exact value of the grass is hard to guess and depends upon the season and the nature of the sward and the management. In the autumn the grass looks as good as in the spring, but it lacks something.

I once turned into young Italian ryegrass in November and the herd's milk yield scarcely responded at all. Yet on the same catch crop the following February the milk yield, even of stale cows, went soaring up.

Once the swallows and house-martins have gone south the quality of the grass usually falls off rapidly. An exceptional season was 1954, when the lack of sunshine had given us grass of rather poor feeding value all summer. But in the autumn the mild, wet weather kept things going and we were still getting maintenance and a good 2 gallons of milk from grass late into November. In very dry summers like 1947, '49 and '55, cattle, will continue to 'do' fairly well on water (if they get plenty of it) and grass that seems almost non-existent. But to get good yields of milk in such times of drought some form of green food as a supplement to the pastures is desirable. And there is no doubt that cattle are healthiest in summers which are neither excessively wet nor excessively dry.

But to return to the breeding of calves. I seldom get trouble from cattle that will not breed. On the rare occasions when I get a home-bred heifer that is abnormal, e.g. one in 1954 which bulled every 11 days—I cull it. I reckon a heifer should be fit and sexually normal, no matter what excuses can be made for cows. As a matter of fact my cows give very little trouble. I have had two bought cattle, one a pedigree Ayrshire heifer and one a Guernsey × Shorthorn cow both of which failed to get in calf with repeated services. Each was dry or nearly dry before I got them in calf. But thereafter I had no more trouble from them. The beneficial effect of a healthy soil had got through to them by then and they remained regular breeders thereafter for many years.

Assuming that the calf has been successfully conceived it is normally born without help being given to the mother. We go and look every now and then if it is daytime, but I don't interfere if everything is normal. Calving is almost invariably outside.

The calf is allowed to suck its mother for 2 to 4 days to ensure a good intake of colostrum, the yellow, vitamin-rich, first milk which sets the calf up in life with resistance to disease and so forth. If the weather is mild the calf runs with its mother for these first few days, coming up with her when she is milked twice daily and sucking her whenever it likes in between milkings.

The calf is then weaned and put in a shed and has to learn to drink from a bucket. For about a fortnight it gets its own mother's milk or that of a cow which calved at the same time, freshly milked and still warm. Thereafter it gets warm milk, but not necessarily that of a newly calved cow. It will get about a gallon a day in two feeds, not less than a gallon anyway. This goes on for about 2 months then it is gradually cut down. So each calf gets from 70 to 80 gallons of fresh whole milk to start it off in life. I believe this is one reason why my cows last so long, going on breeding regularly and milking well into old age.

During this $2\frac{1}{2}$ months the calf learns to eat hay and meal and probably kale or fodder beet tops or silage. (Actually I no longer seem to have time to make silage since I took to growing fodder beet.) So when it is weaned the second time (i.e. from the bucket) it suffers no check to its development.

I have tried hay racks and calf hay nets. Now, however, I simply throw the hay on the floor of the calf box against a wall on clean straw. In this way they waste very little, and if fed three times a day they get plenty of hay even if it is not 'on tap' all the time. In racks and nets they pull out hay, drop it, tread on it and pull out some more.

There are two things I have not yet tried, but which I hope to try soon. One is feeding sea-weed meal in the hope that the extra ration of iodine in the seaweed will keep the calves free from ringworm. The other thing is cud-inoculation—to give the calves at an early age the right flora in their rumens to enable them to deal with hay and such coarse fodders.

The meal the calves get is usually plain oats, ground fairly fine in a hammer mill and with a touch of Gromax Whale meal or Vitamealo added to it. But they may be lucky and get some beanmeal in with the oatmeal. And if we are out of home-grown oats, e.g. in late summer or when the threshing machine has been delayed in coming to us, then the calves get the same Coarse Dairy Ration (a mixture of 'straights' and minerals) which the cows get under the same circumstances. Failing that they get compound calf cake. But I dislike compound cakes for cattle of any age. Cows fed on cake are more bad-tempered and liable to kick than cows fed on oats and beans.

After it is about 9 to 12 months old there is, I believe, no point in feeding a calf on concentrates. Ruminants like cattle and sheep are specially adapted by nature to a diet of coarse, bulky fodders and I am sure we should encourage them in this. Pigs and poultry need all the coarse grain we can spare from milking cows. So the less dependent a cow is on concentrates the better. There is also the point that bulky foods can be produced in large quantities on most farms and in most cases cheaper milk will be produced from such foods than from concentrates even when these are also home-grown. To develop the kind of digestive system that can deal with large amounts of bulky foods I believe a heifer should be fed on bulky foods from an early age. This is not to say that poor quality food should be given them so that their growth becomes stunted, but it means that good hay, some good quality straw, kale, roots, grass are all suitable foods according to the season. I do not feed store cattle on concentrates. Variety of food is helpful. I have seen young calves turn away from really good hay to eat mouldy wheat straw with which they had just been bedded up.

When turning calves out of boxes in the spring I like to do so as the dairy herd or some older young cattle are passing through the yard. Calves freshly turned out are apt to panic on their own. But if older cattle are plodding slowly towards the field the calves usually tag along with them and all is well.

Years ago I had a calf, who, for reasons best known to himself, was called McGinty. His mother was an Irish beef store beast which had 'got into trouble' in Ireland and was not as barren as she was supposed to be. (Incidentally she was the only cow to which I have ever given an orange. She loved it.)

McGinty sucked his mother from birth and by spring time was a lusty youngster. When we turned him and his mother out to grass McGinty abandoned her and set off across the farm as though the devil was behind him and paradise in front. Actually we were behind—a long way behind.

Through hedges and over ditches went McGinty and we followed while the pathetic bellows of his parent grew fainter behind us.

At the farm's southern boundary the calf hesitated and we almost caught up with him. But he was not done yet. He turned about and drove northwards bald-headed through the fainting ranks of his pursuers.

He passed his mother so fast that she hardly had time to recognize him, much less to remonstrate with him. Negotiating gates, fences, ditches and open fields apparently with equal ease he was soon leading by nearly two fields.

What stopped him was the Wadley Brook, the main drainage ditch hereabouts. When we arrived, very weary, one behind the other and armed with a rope halter, McGinty was up to his shoulder in black, muddy water and wearing an expression rather like that of Napoleon on H.M.S. *Bellerophon*.

While two men pulled on the halter I pushed from behind, in the mud, and we eventually got him up the 8-ft. high bank and dragged him home to his dam. And I had a bath before the mud dried.

This sort of thing is usually avoided if there are several calves, and several cows to steady them down.

In summer it is important to see that the young cattle get enough grass, with changes of pasture from one field to another when necessary.

In winter I try to give the bigger beasts, 1½ to 2½ years old, hay for one feed and straw for one feed in two feeds daily, e.g. hay in the morning and straw at night. Chaff (oats, wheat or barley) can take the place of straw. By 'straw', I mean oats, barley or dredge corn straw. Grown organically it is far more palatable for stock than the stuff which has grown up in a sort of hospital-cum-laboratory environment on a modern mechanized farm. I know, because I get a lot of this sort of straw as waste from a local straw mill. It looks lovely, but I only use it as bedding unless very short of grub. Our own wheat straw, grown organically, makes better bedding for cattle—especially young calves. They eat a lot of it. Wheat straw is always thought of as a hopelessly coarse fodder, and so it is, most of it. But young calves will pick out the soft ears and leaves and reject the hard stalks with their tough, indigestible nodes.

I am always meaning to grow enough kale to be able to give the young cattle a field of it in the middle of the winter. So far I have repeatedly failed for one reason or another. The straw they get is often oats or dredge corn straw which has been harvested in hand stacks (see chapter on harvesting) and is far brighter and better as food than straw harvested by stooking, and, of course, far better than the weathered stuff picked up behind a combine.

I have seen out-wintering cattle leave good hay to eat such hand-stacked oat straw, the colour of which is a pleasure to see.

If possible I put the younger stores (those which we cannot accommodate in the buildings but which are really too small to stand up to the bigger heifers) in a yard or sheltered field by themselves. It rather depends what is available for them. Such cattle, 9 to 18 months old, will thrive out of doors in the rain and snow provided they have three things. These are somewhere reasonably dry to lie down, shelter from cold winds and plenty of food. If they have these three things rain only serves to give them a better bloom on their coats and they compare favourably with any of their contemporaries kept under cover, provided the feeding is the same in both cases.

Some green food every day is a great help for such youngsters, and is even more important for them than for the older heifers.

When a heifer is a few months gone in calf, if it is winter time I like to run her with the dairy herd to ensure that she is really well fed, and also so that she will get used to seeing the other cattle milked. Later on, about 2 months before she calves, she can be got into the 'parlour' after morning milking every day and given a little corn. In this way she gets used to the noise of the milking parlour and begins to regard the place in a favourable light. All heifers also learn their names and many will come when called to be milked, which saves the cowman's time.[1]

My cattle are trained from their first milking as newly calved heifers to milk quickly according to the Petersen theory. If a

[1] Many of the cows are now 'house-trained' and will usually avoid dunging in the milking parlour.

cow's udder is very dirty she is hosed with cold water, but is not washed with hot water until about a minute before the teat-cups can be put on. In this way the younger cattle, which never knew any other method of being milked, respond wonderfully, as most of the older ones did too. Quick milking, is, I believe, also partly hereditary. But I never met a good cow yet that could be milked in three minutes.

The amount of corn fed can be gradually increased as the heifer approaches her time to calve. But I do not use the extreme methods of 'steaming-up' that some believe in. No doubt it is very profitable. But I cannot believe it is good for the cows or their longevity, and when it results in pre-milking being necessary it deprives the calf of its proper allowance of colostrum or first milk, which is likely to give it a bad start in life and possibly lead to its early death from some disease like pneumonia which it would otherwise have resisted.

Where docility can be combined with milk yield it is a very desirable thing—especially where the breeding of the bull is concerned. I have been tossed by an Ayrshire bull (whose sire gored *his* owner) and I will not pretend to share Mr. George Henderson's indifference to being tossed. But then I had been pinned against a wall for what seemed rather a long time before he tossed me. So perhaps I am prejudiced.

However, most people will agree that unwillingness to toss is no serious fault in a bull. Docility of bulls seems to go with docility of their female relatives and vice versa.

A bull can also to some extent be kept in a good temper by good management. My bull does not get the exercise he ought to have, but he is chained up where he can often see the cows, calves, people and so on. So he is less likely to be bored, and sometimes I tether him on a chain in a field.

But however docile a bull may be, never trust him. It is foolhardy to handle adult bulls unless two people do it together. This is constantly demonstrated by tragedies. Only the hardiest fools (like myself) live to learn.

To keep the bull healthy we give him kale, grass, fodder beet, etc., as available, in addition to his hay and straw rations. And

13. A very heavy crop of dredge corn. Girl's height 3 ft. 9 in.

14. Another good crop of dredge.

15. A good crop of wheat.

16. A job two boys will think worth doing—drilling spring corn. (The author's eldest son and his friend.)

when he is tied up again after serving a cow he gets meal or fodder beet as a treat to take his mind off the fact that he is again chained up.

The breed of cattle I favour is the Shorthorn. I won't pretend that there are not other good breeds. But I like them and I breed them. Some of my cows are not shorthorns and some of the shorthorns are not dual-purpose cattle at all. This is simply because they are good cows and I won't get rid of them just to make the herd look uniform as to breed. But in the long run all my herd will be Shorthorns.

Friesians and probably other breeds, too, will give plenty of milk and make good beef as well. They are a dual-purpose breed. But I don't like Friesians. They seem to me to be nervous, ill-behaved creatures which panic easily and get everyone handling them much annoyed.

I am breeding my own strain of cow, having scant respect for the generality of pedigree breeders who juggle with their breeding in order to produce a few prize-winning 'plums' which in no way represent the average of their herds. I hope one day to have a herd producing a thousand gallons (or perhaps a little more) of milk per cow per year, each cow calving every year, and a good beef type throughout the herd, so that the old cow, the surplus beast, steer or barren heifer will each make good beef of its class. An old Shorthorn cow will not make prime beef. But an old Ayrshire or Jersey will not make any beef at all—only hide and bones.

Of course we are told by the know-alls (who are often the do-noughts) that there is no such thing as dual-purpose cattle. This extraordinary statement ignores all the beefy cattle of deep-milking strains in various breeds. I have a family of Shorthorns (derived originally from a single bought heifer) where all adult beasts carry a good weight of flesh, and 3-year-old heifers and steers are really fine beef beasts. I don't say they would stand up to comparison with really good Angus or Herefords, but they make good Grade A beef for all that.

Yet the females milk well—900 to 1,400 gallons being about what you can expect per lactation. They, like my other cattle, are regular breeders, seldom returning a second time to the bull. And with all this, and a good size—some are very big cows—they are placid, docile creatures, unbullying and unbullied by the other cows and not easily excited or upset.

This is the family whose type I am trying to stamp on my herd by line breeding. And the results are slowly becoming apparent. They have only one fault and that is low fat and probably S.N.F. in their milk. My next task, after fixing this type, will be to add good butterfat and milk solids and produce triple-purpose cattle as has already been done in Australia with such remarkable success in the famous Darbalara Shorthorns.

Those who advocate single-purpose breeds for Britain are, I believe, hopelessly out of date. We have very little land with which to feed our population and that little gets less every year as the cancer of brick and tarmac spreads over the loveliest country in the world. Beef is already expensive to buy abroad. And as foreign populations grow (e.g. in the Argentine) and industrialization develops in one country after another so beef is going to be harder and harder to buy even if we have the money to spend on it.

It therefore behoves us to grow our own beef as far as we possibly can. Yet we must have milk and we must use a lot of our land to produce food directly for the human population and for other livestock. It follows inevitably from this that we must rely more and more on dual-purpose cattle, making beef as a by-product of our dairy herds. If we do this we can produce our milk and beef with a minimum head of cattle in the country and more land will be set free for other uses. And if we care to use a real beef bull, such as a Hereford, on dual-purpose females the resulting calves will be of far better beef type than calves with Hereford faces and bodies inherited half from skinny, dairy-type mothers. We produce a lot of 'beef' from our dairy herds now. Why not make it good, or at any rate better, beef?

I believe the day will come when we shall not be able to afford to keep cattle in this country which do not combine say thousand-gallon milk-yields with 5 per cent butter fat and good beef type. Of course a few pedigree herds of beef cattle will be kept, especially to breed cattle for export. But the majority of the country's cattle will have to be triple purpose, at least in the dairy-farming areas, i.e. good for beef, milk yield and butter fat.

In 1955 I sold six cows to make room for heifers. They were mostly old cows which had done well in the herd, but two were young ones (5 years old and $6\frac{3}{4}$ years respectively). All were sold freshly calved and all but one were sold without their calves which, being all female, I kept. The average age of the six was almost 10 years. In the open T.T. market they made an average price of $48\frac{1}{3}$ guineas. Not too bad for non-pedigree cows of that age.

In a herd of truly dual-purpose cattle, steers can often be reared with profit. They form a useful 'cushion' between the dairy herd and a shortage of fodder. While food is abundant they will help to control the grazing and manure the land. But when the occasional bad season comes along they can be sold off, whether fat or as young stores, and so ease the food crisis for the females. Without a few steers the crisis would either mean that some valuable females must be sold or else that fodder must be bought to tide them through to better times. Even if the steers are sold at little profit or a slight loss, it is better so than having to buy hay or sell your heifers, and in many years the steers can be sold fat or as stores very profitably.

On the subject of line-breeding, one of the most obvious bad aspects of artificial insemination is the shortening of the useful life of bulls which it is said to cause. Many of the finest sires in every breed are now bought up for use in A.I. centres and their semen is then dissipated over a vast number of cows in a great number of herds.

This use of an outstanding sire on mediocre cows will result in only a fractional improvement in performance of the next generation of females in the herds concerned.

If however the bull had been kept at home and used on his daughters and then on his granddaughters and even his great-granddaughters a very marked improvement might have been expected in the great-great-granddaughters so produced, for they would owe the greater part of their hereditary make-up to the bull concerned. His type would be indelibly stamped on the herd.

As it is, however, the bull will be worn out at the end of three or four years and cannot hope to be used for line-breeding over three or four generations.

This is probably far from being the most serious objection to A.I. but it is one I have not seen mentioned elsewhere.

I have dealt in another chapter with cattle diseases. But I must here mention Bloat, Hoven or Blown or whatever your local name for it is. This is not so much a disease as an acute form of indigestion.

Years ago I lost a beast from blowing on a predominantly red clover ley. Since then I have only had two cases which were serious. One was a beast which was being treated with sulphanilamide and got blown on hay. The other got slightly blown after eating something poisonous, probably out of one of the hedges. This is rather significant for in one case the chemical and in the other case the poisonous plant or what not must have killed off a lot of the 'flora' of micro-organisms in the rumen of the animal concerned. These micro-organisms are essential for proper digestion and their sudden reduction in numbers can apparently cause bloat.

In passing it is worth noting that these micro-organisms of the ruminant stomach have been bred for more than a year in the laboratory on a diet of pure cellulose (filter paper). It is therefore presumable that they must obtain the nitrogen for their body proteins from the air. It appears that numbers of these creatures are digested by the cow in which they live and so the cow apparently gets some of its protein food by nitrogen fixation in its own rumen. If so, this is just one more way in which atmospheric nitrogen is 'fixed' by nature and helps to increase the amount of useful nitrogen on the farm.

Apart from those two exceptional cases of blowing I have never had even a moderately serious case since that first loss I mentioned. One reason for this may be the herbs in most of my leys. Another may be the fact that I do not use artificial nitrogen on my leys to force an unwholesome growth of grass which cattle certainly find unpalatable and may find indigestible.

But I believe the most important reason for my avoiding this trouble is the way I graze my leys. I do *not* turn the cows in for an hour a day or anything like that. I do this with kale when feeding it off in the winter. But the cattle are glad enough by then to go into another field and get straw and water and somewhere to lie down.

But they are never glad to be turned out of a new and tasty pasture in summer time. If they are turned out after an hour or so each day they soon get to learn this. The result is that when they go in they eat recklessly for the whole of their time and they eat much too fast. They pause neither to drink nor chew the cud. And they probably manage to overeat themselves pretty badly. Bloat may easily result and often does by all accounts. The beast I lost was being let in and fetched out of the ley every day.

A much better plan is to turn the cows in and let them stop there. Divide the field up with an electric wire if you can, to give them several small paddocks.[1] This makes for more efficient use of the grass. But let them stop once they have gone in. They will feel under no obligation to gorge themselves and will stop eating and chew the cud when they reach the appropriate state of fullness, continuing to graze again when ready to do so and not before.

Those unfortunate enough to suffer from beasts getting blown might try drenching with soil shaken up in a bottle of water. Mrs. K. M. Studholme tells me that this works in Kenya, so it probably will here.

This system of grazing has one other point in its favour to my mind. The dung and urine of the grazing beast fall where they are wanted—on the ley. We talk about grazing leys as being a wonderful restorative of soil fertility. But this will not be so to anything like the maximum extent if the dung and urine mostly fall elsewhere.

[1] Or strip graze.

CHAPTER V

THE PIGS

I used to have a herd of pedigree Large Black Pigs. This was wiped out, during the war, by one of the more worthless employees who at that time passed for farm workers, who knowingly fed one bucketful of raw swill to a sow and pigs, thereby giving the whole herd swine fever.

The pigs died one or two at a time during a July heatwave. We put the bodies on carts to keep other pigs from getting at them, and duly reported to the police. They reported (I suspect by runner) to the Ministry of Agriculture at St. Anne's-on-Sea in Lancashire, and they perused their records. Then they told their vets in Reading to come and post-mortem the dead pigs.

About three days after we had reported the death, my local vet would be phoned from Reading and asked to do the post-mortem as it was nearly 5 p.m. and no one would be able to get over from Reading that day.

After three days and nights at a temperature ranging from about 70° F. to about 110° F. the inside of the pig was a mass of evil-smelling and putrescent liquid. Diagnosis was impossible.

When more than half the herd had perished a vet. from Reading came over and slaughtered an ailing pig, with easy diagnosis of swine fever. I sent all the rest of the still healthy pigs to the bacon factory under a casualty licence.

My first-born son was a very new baby while all this was going on, and we were rather anxious for his health while the farmyard perpetually contained decomposing corpses and plenty of flies. Perhaps we owed his survival to the large population of house martins which build each year under the eaves of our stable and farmhouse roofs. (Although their principal food seems to be ladybirds they probably eat a lot of flies.)

One good thing the swine fever did was to teach me a lot about the subsoil on my land. We were perpetually digging graves, for hardly was one pig buried in lime when another would be ready for the grave. I dug some of the graves in the rickyard near at hand, but some in outlying fields simply in order to see what was under the top soil.

Since then I have never gone in for pedigree pigs, though I may do so again one day.

Before the war my Large Blacks were fed, apart from grass, largely on bought meal. For a time at the beginning of the war I had the experience of having a herd of pigs which could not be fed, sold or slaughtered, and I have never forgotten it. Never again am I going to rely to any extent on bought meals. Fish meal or its equivalent, yes, but not for the main bulk of the rations. War, politics or international finance could easily cut off the supplies of imported feeding stuffs, but it would scarcely affect my pigs which are fed on swill, home-grown grain and fodder beet.

The fantastically cheap wheat available before the war led me to have a mixture made for my pigs which was based largely on wheat and wheat products. This resulted in the pigs 'going off their legs', a trouble which was put right by changing to a more orthodox mixture. Perhaps cooked wheat would have been all right. I don't know.

A high proportion of oats in the pig meal is perfectly good so long as the oats are a fair sample and not all husk, and they are safe. In fact a lot of oats in the meal is probably better for the health of the pigs than a lot of barley, and it is quite possible it would give a better grading result than the very fattening barley meal.

I consider an all-meal diet for pigs quite the wrong way to feed these creatures. The pig is an omnivore. It has the most complete and comprehensive set of teeth (44 of them) of any farm animal, and its teeth are adapted for eating *everything*.

They are something like our teeth only much, much more so. The teeth of an animal give a very good clue as to how nature intends that animal to feed. It is as stupid to feed a pig on nothing but finely ground meal as to expect a dog to thrive on nothing but oatmeal or boiled cabbage.

The more varied a pig's diet is, the happier and healthier he will be. To try to provide this variety by putting penicillin in his pigmeal is about as reasonable as feeding people on white bread plus laxatives to cancel the effects of the white bread. And it means that a lot of skilled laboratory workers are needed to fatten pigs!

I cannot claim to have done extensive and carefully controlled experiments with the feeding of antibiotics to pigs. But the little I have done has convinced me that creep-feed pellets plus antibiotic are inferior to fishy swill as a food for baby pigs. The pigs would certainly agree with me.

A pig's love of an omnivorous diet can be embarrassing at times. I have had more than one pig which took to eating poultry. The hens often come round the troughs to feed on the pigs' food, and this gives the pigs their chance. All you find is a wing or two and a few feathers mixed up with the bedding straw. Such pigs get sold quickly.

A less serious affair was the litter which kept a nest in the corner of their sty for a pullet to lay in—and then collected the egg! I watched the piglets one day playing with each other in the sty. Suddenly one broke off from the game and walked over to the sitting pullet. As cool as you could wish he stuck his nose under the bird, lifted her up, saw there was no egg yet and put her down and went back to his game.

A pig that is fed on swill probably gets all the antibiotics known to science and several that are unknown,[1] for Penicilliums (of various species) are among the commonest mildews on stale food. (Or does the boiling destroy their virtue?) This sort of thing is a pig's natural diet—a bit of everything. And I

[1] There is apparently a species of yeast which sometimes affects the newly boiled swill, causing it to ferment. This probably supplies valuable vitamins of the B complex.

consider the real function of a pig on a farm to be the conversion of refuse to meat and manure.

However, the need for home-grown meat and bacon to help the nation's house-keeping means that we must keep more pigs on some farms than can be fed on scraps. Hence the need for dredge corn, feeding barley and fodder beet.

Swill for pig feeding has, by law, to be cooked for at least an hour before it is fed. This is a necessary evil to check the spread of swine fever, foot and mouth disease, and so on. This means that pigs so fed must have some raw food if they are to keep healthy for long. My pigs get uncooked, home-grown grain, and also green stuff such as kale or fodder beet tops, the roots of fodder beet and, sometimes, raw potatoes. They also get a small amount of uncooked scraps from my household, my wife being scrupulously careful to keep it free from contact of any sort with meat or bones. Occasionally I give them uncooked herrings' heads from the fishmonger. And they get all edible waste from the garden.

Swill is low in protein, and pigs fed entirely on swill with a small health ration of green stuff make very slow growth. I have found that this state of affairs can be put right by adding to the swill fish and butcher's scraps and poultry offal. This stuff is cooked in with the swill, and the fish bones, at least, are softened by being boiled with the vegetable peelings, etc., till they are often as soft as biscuits and can be crumbled in the fingers.

During the outbreak of myxomatosis in 1954 I boiled up a lot of rabbits in the swill as part of the pigs' protein and mineral ration. And when we kill a lot of rats and mice at threshing time they go into the swill. I once boiled a fox. The dogs refused to eat it. I think in the end the pigs ate it, but it wasn't popular with them, and the smell of fox is worse (otherwise unchanged) while it is cooking! Nowadays foxes go into the compost heaps—whenever I can manage to kill them. I am told that in Spain pigs eat snakes. Pigs free to root undoubtedly get a lot of protein and minerals by eating earthworms. Anything like this which can save a little expense on fish meal is welcome.

The addition of fish to the swill supplies all the proteins and minerals a pig can require and makes a perfect food for pigs of all ages from three weeks old upwards. When there is not enough fish to balance up we add white fish meal, whale meal or a proprietary protein-mineral supplement after the swill is cooked.

Pigs fed on swill with a liberal protein content grow like mad. It is impossible to say how much protein there is in my swill, for it is a wet food, even before cooking, and the amount of fish added varies from day to day. For the same reason I cannot give figures for food consumed against live-weight gains. But I have compared the growth rates of my pigs with those of published figures for pigs fed on meal and antibiotics or alphabetical 'factors', and my pigs seem to do as well or better.

If live weight is plotted graphically (upwards) against age (horizontally) allowing one small square of the paper for 1 lb. live-weight and one small square for each day, then a growth line rising at 45° means that the pig is putting on 1 lb. per day. If the rate of growth is faster the angle will be steeper. Hence the angle of the growth line above the horizontal forms a quick way of estimating the rate of growth of the pigs and comparing it with those of other pigs.

I have had growth lines rising at 55°, 60° and even 65° which seems to compare favourably with most other figures I can get hold of.

A point to notice is that when a pig gets to about three to four weeks from slaughter it should not be fed any more fishy swill. I put mine on to swill and Vitamealo or whale meal or any other protein food that does not contain more than a small proportion of fish. This avoids the risk of fishy bacon.

Swill, especially if it contains much poultry offal, is such a fattening food that my present practice is to cut out *all* swill feeding for the last two months or so before the pigs go off for bacon, keeping them on meal and fodder beet from then on.

In theory this will give me better grading results and more profitable pig production. But in practice there is a marked slowing up in the rate of growth of the pigs when they are deprived of the swill. This is probably just what is wanted to make the grade better. But if too long a delay is produced in getting them off to the bacon factory a reduction of profit will result from slower turnover and longer occupation of buildings and use of labour for a longer period.

Perhaps the solution is to go on slapping the swill into them, but without fish in it, and sell them off as porkers.

My pigs are very seldom ill. This suggests that if people ate at least some of the peel they throw into the pig bin they would be equally healthy.

I suppose most of the food in the swill has been grown on artificial manures, and there is usually a fair proportion of white bread in it, and it is cooked (a lot of it twice) for at least an hour, yet a little organically grown grain and some organically grown green stuff as supplements to the swill are enough to maintain blooming health almost without exception.

There are, of course, two fallacies in this argument. Mankind is given to worrying and this is a potent source of ill-health. And mankind breeds from the unfit, whereas weakly livestock are almost always culled (except in pedigree herds).

In most breeds of livestock there are good and bad strains, and this applies to pigs. The most important things to look for in any strain of any breed are: good type for bacon or pork; good, quick, thrifty growth; good milking qualities in the females; and fairly large litters, the number of pigs reared being more important than the number born.

The best way to judge milking qualities is by the weights of the piglets at weaning. It is often stated that a sow's yield falls off rapidly after the piglets are three weeks old, and that, to test her milking capacity, the litter should be weighed at three weeks. The weighing at eight weeks, when the piglets are weaned, is then a measure of your efficiency in creep-feeding the litter. While that is true in part of the eighth-week weighing, I do not agree that a sow's milk yield drops inevitably after three weeks. No doubt it does so if she is wrongly fed, i.e. fed simply on an extra large ration of meal, or if she simply does not get enough food. My sows get swill, meal, kale, grass, fodder beet, etc., and plenty to drink, and it is quite obvious that they go on giving abundant milk. Not only is this shown by the keenness of the piglets to suck the sow and their persistence in doing so, but also by the way she loses flesh. At three weeks she may still be quite fat, but at eight she will be skin and bone despite three feeds a day. Then, when the piglets are taken away, her udder at once becomes distended with milk even though her liquid intake and her ration of food are cut down at weaning. After weaning she soon fattens again.

A gilt in 1955 which had only four pigs gave so much milk that the piglets only started to feed properly at eight weeks. Up till then we never saw them eat anything except a bit of green stuff (chiefly the big, white-flowered convolvulus, growing near a pond) and whatever they could get by rooting—probably worms. Accordingly we left them with their mother for more than eight weeks to avoid a check at weaning.

Weighing the pigs is a thing that should, ideally, be kept up weekly from weaning to bacon weight. Even where weekly weighing is impossible a fairly frequent use of the scales makes for efficient management. In fact I would describe weekly weighing of pigs as being as important as daily recording of milk yields of dairy cows.

Some people claim to produce weaners of 45 lb. or more *on an average*. I suspect that they wean at nine or ten weeks after birth, or else that they have mixed up the terms 'average' and 'best'. My weaners seem to average about 36 lb. to 38 lb. at eight weeks old, though individuals may scale 40 lb. or more. Since this is well above the national average it may be helpful to describe my methods.

As soon as the baby pigs venture outside their mother's sty—by means of the 'creep' or pophole—they are offered a slop of oatmeal and water. This is usually about two weeks after their birth. If the sow is milking well, the piglets seldom take much interest in the food until they are about three weeks old. But we offer it every day or two in case they will try it.

As soon as they begin to eat the oat slop they are given it regularly, and it is soon changed for swill or swill and meal, the swill being rich in boiled fish offal. This is very popular, and the piglets soon come to regard swill as the only food fit for them, squeaking and grunting with disgust if they are offered bought sow-and-weaner meal or even home-made pig meal. But if swill is short they make do on meal.

Pigs of four weeks old and even less will nibble fodder beet if they can get it. When they have got used to it early in life in this way there is no trouble starting them on it later on.

Feeding them in the open farmyard is tricky because geese and chickens are apt to steal the food in the early days. But we manage to see fair play and feed them as often as possible (two or three times a day or more) until they are old enough to chase the geese and hens off for themselves. Or we feed them in another, empty sty, leaving the door fixed ajar or with a hole in it or a gap under it which admits the piglets and excludes the geese. Pigs have sharp noses and sharp wits and it is easy to teach them where to go for their food in this way.

'Creeping' into the farmyard has great advantages over any confined space. The piglets get more exercise, more fun (and give us more) and they can go and rout in odd patches of soil or dung heaps, or they can graze grass or kale, or steal some fodder beet. They can drink as much water as they like when they like. And the exercise they take means that we do not get those excessively fat weaners.

Up to weaning time, assuming that water is available to them in the yard (or the sty), their creep feed need not be sloppy but fairly stiff. After weaning, when their mother's milk is denied them and they are shut up away from water (unless it is laid on to the sty), they should be given a much sloppier mash. This can generally be thickened as the baconers grow older.

At five weeks or thereabouts the boars are castrated. I do this myself with a sharp, sterilized scalpel or dissecting knife. It is a help if the person holding the piglet holds it head-up, with its hind and fore feet gathered up together in his hands. Should there be any swelling of the cuts following castration, causing a piglet to be off colour, sit it for several minutes once or twice a day in a bucket of hot, strong Epsom-salt solution. It is very rare for us to have to do this. For a few days after cutting them we are more than ever careful to keep their sty well bedded up with clean straw. This is important, too, just after weaning, when they miss the sow as a hot water bottle.

At about eight weeks or just before, they usually find a way into my vegetable garden. If they become a real nuisance they are weighed and weaned and ear-marked (unless their markings are very distinctive) so that at future weighings the progress of each pig can be checked, and not just the average of the whole lot. In any case they are weaned at eight weeks. Marking each pig is helpful when they approach bacon weight, for it enables the heaviest pigs to be picked out without error, which is not so easy by eye as might be supposed.

On this question of weighing I may say that I have no expensive, pig-weighing machine. I simply put each pig separately into a box of known weight (I check the weight every weigh day) and lift it on to the corn scales. With small pigs it is easier to weigh yourself and add to the scales any convenient thing such as stones, bricks or fodder beet to make your weight up to a round figure. Then grab the pigs one at a time and stand on the scales and deduct twelve stone or whatever it is to get the weight of the pig.

Towards the end of their time when they are seven score or so in weight it is quite a struggle getting them into the box and then lifting them on to the scales. So I am often content then to weigh one or two pigs only in each batch, at least on alternate weigh days. I choose, usually, one that is exceptionally heavy for the batch (for fear he gets over weight and I lose money on him) and one which previous experience has shown to be usually about the average of the bunch. If any unforeseen results appear (if, for example, no gain has been made during the week) then we weigh the rest and see how they have done. Or you can weigh again after three or four days.

Meanwhile how has the sow been fed? I do not believe in all-meal feeding for a milking sow. It is unnatural. It is bad dairy farming to feed a cow entirely on dry food--meal or cake and hay. You get more milk and healthier cows if grass, roots or silage are included in the ration. I am sure the same applies to pigs. As well as swill and meal my sows get all sorts of odds and ends including greenstuff and fodder beet. I am sure this varied diet is the right way to keep them healthy and milking well, and to keep the milk rich in minerals and vitamins. Certainly I have never had a case of piglet anaemia in a home-reared litter. A milking sow should be allowed as much water in a day as she will drink. If water is not laid on to the sty, give her as much as she wants at the same time that you give the third feed of fodder beet. Or after a feed of meal and water offer her plain water some time before the next feed. In hot, summer weather this is particularly important, and we do it with pigs other than milking sows in heatwaves.

There are two small but interesting results of swill feeding. One is that the pigs never get lice. I think the explanation is that the greasy swill gets behind their ears and kills the lice or their eggs just as oil would do. This is confirmed by one exceptional pig that got lice in 1955. She was a gilt living in a box by herself and was the complete lady. With no competition at meal times she ate with great daintiness and never got swill behind her ears. I had to put oil there instead. She is a most fastidious feeder but a good 'doer'. It is almost impossible to keep her thin. She was reared on a foster-mother.

And swill-fed pigs never suffer from worms. Probably this is just because swill keeps them in such thriving condition.

I usually feed my pigs three times a day up to bacon weights. Breeding stock above that weight get fed only twice, or perhaps only once, getting water alone for the second feed if they are getting too fat. Or they may get a minute token feed at the second or third feeding to keep them quiet if they hear their neighbours eating. As soon as a sow farrows she goes back to three meals a day.

My pigman feeds at 7 a.m. and again about 3.30 p.m. or thereabouts, earlier if we are hoping to go haymaking or harvesting. And then I usually give the last feed myself between seven and eight at night. Electric light in the yard and buildings is a great help in this business during the winter months.

I sometimes buy pigs in the market which look healthy but have been very poorly done. They may be skin and bone, but if their eyes are bright and they do not appear actually diseased they are often a very profitable buy. Such pigs, contrary to popular belief, often turn out very good 'doers' when they get a chance on good food. And they can often be bought very cheap. Some of my best pens of fattening pigs have been bought like this as very rough-looking stores. But of course it is much preferable, I think, to breed your own weaners. They seldom look back from birth onwards, and you should reap the profit of the pig-breeder and that of the pig-fattener. If you also grow your own food or most of it you should reap a further profit—that of the crop-grower.

Another common fallacy is that the dillon or runt of the litter (the pig, if there is one, which is far smaller than his mates) is always a poor doer. I have had small members of litters which, given the chance, have grown faster than their big brothers and sisters (steeper growth lines).

When I get pigs home from market I weigh them at once and calculate what I gave per pound live weight for them. If weaners of 30 lb. or so live weight were averaging about four shillings per pound or more, and mine, looking a bit rough, cost me three shillings per pound or less, I am pretty well pleased.

Of course it may happen that one of these pigs will die. But even when the cost of the dead one has been shared between the others, their price per pound often works out very cheap.

When a bought pig dies suddenly it is as well to have it properly post-mortemed by a vet. For one thing it makes you very safe with the police if they come to see the pig during its four weeks' quarantine.

I remember one pig of mine, a home-reared one actually, that died mysteriously many years ago before my present firm of vets came to the district. I got a vet to post-mortem it. His explanation of why it had died has often made me laugh. It had, he said, a small gravelstone in its stomach! Not very remarkable for an animal that routs in the soil for its food if given half a chance, and that will eat clinkers and small coal for their mineral content! He was said to be a wonderful vet with horses.

Nowadays I seldom let my sows out of doors. They do such a lot of damage to good pastures by rooting. I have not yet tried them on special pig leys (with chicory and clover) sown conveniently near the buildings. In theory ringing their noses stops rooting. In practice, rooting, in fairly heavy soil, removes the rings from their noses. So my sows and older stores and baconers are confined to boxes (sties) or to narrowly defined electrically fenced areas with pig arks or improvised shelters.

My Large Blacks used to run loose on the farm in the days when I had more grass (much of it poor) and less arable. They hardly ever visited the clay land grass fields with their poorer swards, but when the acorns were ripe they knew, possibly by their sense of smell, and would cross two or three fields to feed under the big oak trees. They used to vary the very constipating acorn diet with laxative grass and crab apples, and so took no harm from them.

The good sense of pigs is in marked contrast to the mistakes sometimes made by cattle and horses which will gorge on poisonous foods and kill themselves. Pigs will avoid unwholesome foods unless confined to a sty and starved into eating them. It never pays to force a pig to eat any particular food. It knows best what is good for it. That is not to say that you can't coax a pig or pigs to start on fodder beet by keeping them a little short of other food. In that case you know the food is wholesome.

Pigs not only show this good sense about what food to eat, but show also intelligence superior to other farm animals. Quite young pigs will learn that an electric fence is only dangerous when the machine is ticking, and will listen for the machine and act accordingly. Yet I never knew dogs, horses or cattle to learn this. A friend of mine who keeps a great many pigs has one that can tell if the wire is alive by *smelling* it!

The pig's fondness for some fungi, including truffles, is well-known. Yet they refuse to eat horse mushrooms either young or old.

The shortage of pig foods during the war produced some most

astonishing suggestions for emergency feeding of pigs. One such, made I believe over the wireless, was that we should grow sugar beet for pigs. For a farmer already equipped for this crop this was probably a good idea. For those not so equipped it was madness. I fell for the idea and grew about half an acre. The roots could not be pulled by hand. They were bad enough to get up with a fork. And the field I grew them in, though it is now one of the driest on the farm, was then very badly drained, and the soil was heavy. We almost got stuck there beyond help or hope. In the end we floated a couple of pig arks on to the mud and let the pigs do their own digging. The joy about the modern fodder beet is that they yield much better and are easy to lift.

Another bright idea was that we should feed horse chestnuts to them. We could scarcely believe our good fortune. How amazing that such a wonderful source of food should so long have been neglected! My newly wedded wife and I spent many hours together collecting them from trees belonging to friends. Then we tried the pigs on them. They ate about one each raw and two or three when boiled. After that they absolutely refused to touch them.

The French, we learned later, had used horse chestnuts during the occupation to make home-made soap. Years later I found a big sack of very old and dead chestnuts and put them into a compost heap. What a waste of potential soap! Or was that recipe like the one for pig food?

Pigs should always be well bedded up. Not only does plenty of litter mean plenty of pig manure with minimum loss of the valuable urine, but it makes for thrifty pigs. A pig shivering on wet dung or lying on cold concrete must use a lot of his food to keep warm and correspondingly little for growth. If he can crawl right out of sight under his bedding, preferably among a heap of his fellows, he is really happy and the loss of heat from his body is reduced. It is amusing to watch pigs in a large pile. The top ones grow cold, wake up and crawl in under the bottom of the pile with sleepy protests from the others. Then the new top layer gets cold, and so on, with a sort of convection current of pigs. They sleep more peacefully if the top ones are under straw and don't get cold. With plenty of bedding there is no need for airspaces under the floor or special dunging passages.

During heatwaves less bedding should be given, and the pigs will then wallow in the wet dung. Heatwaves are rare in England and this need not much reduce the amount of dung they make.

A sow that is about to farrow should not be bedded up with straw unless it is cut into chaff. Wheat chaff (glumes off the ears) makes the best bedding. It is dry and warm, and the piglets cannot get tangled in it and are therefore less likely to get squashed by the sow. Failing chaff, use sawdust, but it is rather choking stuff for new-born pigs. It is excellent with chaff at about a week old.

A sow and her litter during six months (the litter only being with her for two months and running outside during most of the daytime) will make about 6 tons of dung if liberally bedded up. They could probably be made to make still more with even more generous use of bedding.

I do not know how much straw goes into this 6 tons. We put a fresh pitch or two of bedding in every day (or we like to think we do) when once the piglets are strong enough to avoid getting tangled up in it. This is probably at about three weeks, or earlier if, instead of straw, cavings can follow chaff.

This means that two sows in a year, plus their pigs up to weaning age, will make about 24 tons of first-class dung. Enough to give a good dressing to an acre. If each dressing lasts for, say, three or four years' rotational cropping, two sows rearing weaners will make enough dung for three or four acres of land. Minerals may be needed to supplement those in the dung, and I assume that more humus will be supplied by leys or catch crops. If the land were in grass all the time, and being grazed most of the time, the dung alone would probably maintain or increase the fertility of the soil.

I have a thermostatically controlled, electrically heated foster-mother for use in conjunction with infra-red lamps when a sow has a very large litter. I reckon it is best to leave up to twelve pigs with a sow. But if she has fourteen to twenty, as some sows are apt to do, it is probably wise to take away all the smallest, leaving the sow the twelve best piglets to rear herself.

Come what may in the way of casualties, I leave them with her for twenty-four hours to get a dose of colostrum. Then we rear them on cows' milk treated more or less according to the instructions of the makers of the foster-mother.

Actually I add 'Ostermilk' instead of powdered skimmed milk and I use 'Minadex' syrup instead of cod liver oil. It supplies vitamins A and D and a lot of minerals and it mixes with the milk. Cod liver oil merely floats about on top in drops and is very awkward.

We also found that medium-hole baby's bottle teats are needed for the first few days as the piglets (the smallest in the litter) won't or can't manage anything tougher. When they begin tearing these teats or pulling them off, go on to the teats provided by the makers. If they pull these off after a while, try giving them warmed milk at frequent intervals in a shallow trough. One small gilt we reared was a fool at sucking the teats and did very poorly. But offered milk in a trough, she forged ahead well and made a fine pig.

Once they get on to feeding from a trough you can go back to cod liver oil, as it is then no disadvantage to have it floating on top.

Though it is against all my ideas, I must admit that pigs reared thus *do* as well as or better than pigs reared naturally. Whether they are as constitutionally strong I cannot say, but the gilts so reared are likely to make good mothers as they are so quiet and used to being handled.

Where two or three sows can be induced to come in season and be served and farrow together it is very useful. Too many pigs in one litter can be put on to another sow with too few. The trouble is that a small pig keeper may want his sows to farrow at well-spaced intervals so as to keep up a steady flow of pigs from weaning to bacon weight without overcrowding the buildings at one time and leaving them empty at another.

Where this swopping round of parts of litters is done it must be done very soon after birth. Within the first few days of their lives each baby fights for and establishes his right to one and only one of his mother's teats. Thereafter, that is his teat and no one else's. If he dies that part of her udder will go dry. And any teats not sucked soon after the litter is farrowed will dry off. Hence it is no good putting pigs to a sow four or five days after she has farrowed, no matter how few pigs she may have. I once had a gilt with only one baby. She was very conspicuous and was known as 'The Amazon'.

CHAPTER VI

THE OTHER STOCK

I used to keep a variety of other stock besides cattle and pigs. I kept goats, fattening cockerels, geese, laying hens, turkeys, ducks, horses, bees, almost everything, in fact, except sheep. Sheep would be just one more job for the boss on a small farm. And the boss has already more than he can manage.

Nowadays the turkeys, goats and most of the fattening cockerels have gone, the geese are being severely reduced, the horses have become a horse and we have few ducks at the moment. All this has happened in the interests of specialization and greater money profits. I don't think it is a good idea from the point of view of theoretical farming. But it certainly pays better in practice.

However, I may as well record some of my experiences with these other classes of stock.

The goats which we kept supplied milk for the children. The cattle are all 'Attested' and we did not keep goats from any fear of tuberculosis.[1] But goats' milk is supposed to be best for children, and so we kept goats. They are in any case exceedingly clean, fastidious animals, very friendly and lovable. Only the males have an objectionable smell.

We used the goats to utilize odd corners of grass or weeds where we cannot put cattle, such as in rickyards and the corners of fields and so on. They were tethered each day in a fresh place, and so the damage they do when loose was avoided. We also used them for trimming off the invading brambles from our hedges. But care has to be taken that they do not browse the hedge proper. During the night they slept in the stable with the geese, and added their quota to the work of treading straw into 'dung'.

[1] T.B. is said to be rare among goats.

An interesting point is that goats' milk tastes much like cows' milk when they are fed on grass and hay. But when they are eating brambles, thorns, briars, privet and other trees or shrubs (a goat's natural food) their milk acquires the characteristic 'goaty' flavour which can be objectionable if very strong. It also acquires this taint if a billy is on the premises, though why this should be so, I cannot say.

As I say, we kept goats for the sake of the children. Our youngest son is being reared on cow's milk, and I am bound to admit it suits him just as well. So we have got rid of our goats. On a moderate-sized farm like this where cattle can be kept there is no appreciable amount of profit in goats in return for the amount of labour they take (mostly my wife's).

The place of the goat, to my mind, is for the smallholder or the cottager who likes to be independent and have his own supply of home-grown milk, cream and cheese, and some first-class muck for his land—the sort of person who cannot keep a cow because he or she has no field or no shed big enough to house it at night, or because they are too old to handle big stock. For such folk the goat would seem ideal, provided they get them young and train them in ways of gentleness. They can eat a lot of house and garden scraps and (tethered) graze roadsides and lawns and browse on patches of bramble. They can winter on hay gathered from grass verges, and fodder beet grown in the garden. And they will eat things that cows dislike, such as burdocks. A lot of winter 'hay' for goats could be made by cutting whole burdock plants and hanging them up in a shed to dry. They also eat nettles and young, tender thistles more freely than cows will.

Our first goat had been a lady's pet and insisted on being milked in the kitchen. After a time we got bored with sweeping up, twice daily, the showers of dung pellets and thereafter milked her in a shed. She never forgave this, and from then on it took a strong man to hold the goat and a strong woman to do the milking. However most goats are gentle enough.

Kid meat is like the most wonderful lamb you ever tasted, and then a little nicer than that. The trouble is that it takes a hard heart or the compulsion of sheer necessity to kill anything as irresistibly charming as a young kid.

Khaki Campbell ducks used to do well on a farm where there are four ponds in and around the farmyard. When I first came to Pucketty, I remember, I heard frogs croaking at night in the yard. I have never heard this since the ducks arrived!

The Khaki Campbells became unproductive and I got rid of them. While I was wondering whether to get any more, the Muscovies happened. One morning near Christmas time a beautiful, black-marked, white Muscovy duck arrived and clearly indicated that she had taken us on. She might have come from any one of half a dozen farms, for they are strong fliers, so she stayed and was fed. Anyway we could hardly refuse her asylum from the slaughter we guessed was going on wherever she came from.

I cannot bear to see any animal living in spinsterhood. In the spring I bought a drake, clipped one wing and turned him out in the yard with the duck. They each appeared delighted with the other's sealing-wax face and were soon inseparable. In summer she sat on eighteen eggs and hatched and reared fourteen young ones.

So now we have Muscovy ducks. They are elegant, delightful creatures, live on tail wheat and what they can scrounge for themselves, and supply us with some excellent table birds in the persons of the surplus drakes. Foxes are the chief handicap with such independent-minded birds, but we give them what protection we can.

The young Muscovies feed in exactly the same way as the largest whales. These little, egg-shaped lumps of black and yellow fluff steam up and down the pond in the yard with their beaks dipped into the water filtering out the plankton of Daphnia (water fleas) and similar crustacea which are the best possible food for the ducklings.

This pond gets a lot of duck and goose dung in it and some washings of rain water from the farm yard. As a result it grows an incredible amount of microscopic algae which colour the water pea-green in spring. Then follow the water fleas which feed on the algae, etc. These turn the water from green to pink, the colour of their bodies in the mass. Then come the ducks which leave more dung in the pond and so the cycle is complete. I further take advantage of the fertility of this pond to raise young cricket-bat willows there. Slips 18 in. high stuck in the mud in spring will grow more than 6 ft. high by the end of summer. But I am digressing.

A word about killing ducks and geese. They are highly intelligent, sensitive creatures with imaginations of their own. To kill one duck in front of the rest is sheer cruelty to the others, and the same goes for geese. I always try to conceal the death and plucking of ducks and geese from their companions. With chickens it is entirely different. A hen will watch you kill another hen and then walk up to you while you pluck it, and pick over the feathers for juicy young feather stubs to eat. I doubt if a hen can even imagine the next five minutes of its life.

To kill a goose hold its feet in your hands and lay its head on firm ground. Lay an iron bar or stout stick lightly across the nape of its neck, keeping the head chin-down. Put the toe of one foot lightly on one end of the bar. Then put the other foot forward and hold the bar firmly down on the ground. At the same instant pull sharply with your hands. With a little skill you can kill the bird instantly without putting any weight on its neck until the moment that you pull and break it. This method bleeds the birds just as well as knifing their throats and is far more humane, since the bleeding takes place after the bird's neck is broken.

I have cut down on geese. They have a regrettable habit of starting to lay in the early spring when no hens are broody. To rear any number you have really got to have an incubator. Unless you do it in a big way all this wastes too much labour. Now I keep one gander and two or three geese and let them sit on their own eggs and hatch them. Poor mothers get eaten as 'boilers' when all our children are at home and there are plenty of strong teeth available.

The chief virtue of geese is in giving warning of wet weather. When the old birds fly it almost always rains within the next day or two. This is particularly so after a long dry spell. They are far better prophets than the Air Ministry (though I give that body almost full marks for frost-warnings), and they can detect coming rain before there is the slightest fall in the barometer or any change visible in the weather to human eyes.

My geese seem to know the boundaries between my neighbours' land and mine in some strange way. Though they will sometimes stray into an unusual field in search of a change of diet, they do not take kindly to feeding in new fields and have never, so far as I know, been seen to trespass on any other land than mine, though they often go to within a few yards of the boundary fence; and a goose I sold to my neighbour one spring flew home and spent the summer with my flock. Their normal range is the three or four fields round the buildings, to which fields they have access all the year round. Since they graze a bit here and a bit there, I cannot even estimate how much grass they eat compared to, say, a cow. It is certain, however, that their dung is very good for grassland. And they sometimes do a bit of weeding by digging up buttercup corms and eating them.

Goose eggs less than a week old usually hatch pretty well, stale ones very poorly. Failing hatching we eat them or sell them. An egg normally weighs 6–8 oz. So it takes a good appetite to manage one boiled. We once had a double-yolked one weighing 13¾ oz., and it made scrambled egg for the whole family.

Geese ought to pay well if they are run on a big scale in large, grass orchards with bees, or some such arrangement. For they, alone among poultry, can produce meat entirely off grass. They like grass at the short, high-protein stage, 1 to 4 in. long, and are content with this, though they will eat other foods as well if they get a chance or have to do so. Since we took to growing fodder beet our geese have been even more delicious and fat. A poulterer to whom I gave one at Christmas time, in return for fish and poultry offals, declared that it was the best goose he had ever eaten.

Reared under hens, or in incubators, it is perhaps easier to give the goslings a little meal until they are about a week old. But too much can kill them and it is not necessary to give any at all. Chopped dandelion or clites (goose grass) or a grass turf may help to start them off on green food. Hens cannot appreciate a gosling's passion for grass and will keep them pottering about in the farmyard, dry and dusty, while they scratch for food the goslings will not like. It is rather touching to see the patient politeness with which they wait while the hen they take for their mother bores them nearly to death and starves them. The solution is to pen the hens on grass. Then the goslings will graze and graze, until, in the evenings, when they hiss at you you can see the last lot of grass blades stuffed in on top of the rest at the back of their throats. Reared with the geese they are not allowed to eat meal if it is offered, and spend all day on the grass with excursions to the yard for grit or to the pond for drink and washing. They need large quantities of sand as grit, and it should be specially provided for them if they are not running with the old birds. Various weeds and cultivated green food will be acceptable to geese, but *good pasture is all they need from the egg to maturity.* If they are to be killed as green geese at Michaelmas, this is all they should be given unless stubbles are available to finish them on grain that cannot otherwise be used. Make no mistakes about it, grass is the food on which they will grow quickest. For the Christmas market, with grass becoming poorer and scarcer from October or November onwards, a little grain, meal, bread, boiled potatoes, cabbage, kale, fodder beet or apples or some such will help to keep them going and finish them well. Even in winter they will often do a lot of grazing, just as, of course, the wild geese do. A Chinese gander on Emden-Toulouse geese gives a snug little table goose which fattens better than the very big ones do, and is likely to be more popular with small families and small ovens.

Geese are terrible thieves but very noisy ones. They preface all their raids on pig troughs, corn ricks or the open door of the meal house with a frank outburst of cheering which warns you what is afoot. Contrary to popular belief they are uncertain 'watchdogs'. In 1955 a lot of ducks and hens were taken by foxes from under the noses of the geese in a well-lit yard without their uttering a sound.

Geese themselves are almost too much for a fox, and a compact flock with a few savage old birds to guard it will seldom suffer loss, even if left out at night. Stray goslings on their own, however, may get bitten or even carried off at any time, day or night, by foxes, especially in hunting districts where this pest is still half-protected.

These beautiful birds have the reputation of keeping a farm free from red-water disease of cattle, with what truth I cannot say. One thing is certain: they are superbly healthy themselves.

Anyone who has watched little pigs will, I think, agree that they have at least a rudimentary sense of humour. Not so geese. They can jeer but never laugh. Archie was a young gander, incubator-hatched, who grew into a huge bird. He fell madly in love with a goose a year older, Ros by name. The happy pair used to wander about, tête-à-tête, murmuring endearments to each other and as oblivious of the rest of the world as any human lovers.

One day Archie had a fight with the head gander, Mussolini, and gave him a hiding after a battle of which I saw the finish. Archie returned from this triumph, which had astonished him as much as it did me, looking about twice natural size and with his beak pointing straight at the sky. He came galumphing back to the admiring Ros, shrieking the usual boasts of a gander who at least fancies he has been victorious. He couldn't see through his chin and so fell headlong over his beloved, knocking her down and floundering over her back, head down and feet up. I nearly died of laughing, but I can emphatically state that not a goose in the whole flock sniggered.

My laying hens are few in number and kept in a haphazard way—what a friend of mine calls 'the bung system' (or should it have capital letters?). I told this to an earnest young man once and had to explain hastily what I meant. He was obviously on the point of asking me who Dr. Bung was?

Now this learned gentleman's system is perhaps the most widely popular among general farmers where poultry are concerned. It has many advantages over deep-litter, folds, batteries, henyards or anything else that may yet reach us from Harper

Adams. The chief of these advantages are flexibility and labour-saving. Indeed there is only one essential feature common to the best practice of all the leading exponents of the art and that is to do nothing scientific and nothing which takes a lot of trouble. In its barest essentials the system can be summed up in the words of the learned doctor himself: 'bung 'em out near the farmyard and bung some corn into 'em twice a day'.

Needless to say on this exquisitely simple basis quite an elaborate system of poultry-keeping can be built up, the details varying to suit the particular needs of the farm or farmer. In my case eight-week-old pullets are bought about every other year. I always mean to ring them for identification, and never do, so that after a year or two they are indistinguishable from hens of a venerable age. Culling for the pot or for sale therefore takes the form of a lottery with me and my pigman representing chance, or a large part of it.

At first the pullets are carefully tended and shut up at night. To prevent my forgetting to do this I write 'Hens' in chalk on one of the riser boards of the stairs, so that I see it when I go up to bed. Gradually, however, they start roosting in trees, on old carts or on the beams of the old cowshed now converted to pig sties, etc. Then I take the few that are still roosting in the house, chuck them up in the tree with the others and jolly well teach them to look after themselves. Then, like the sailors, they are up aloft and safe from foxes if they do not come down too early in the morning. Once this stage is reached my wife is allowed to wash the stairs.

They lay in the nettles, in the hayloft, in the meal house, under the paraffin tanks, in the garden hedge and in other odd places. Hens' nesting becomes quite an art and is a pastime which everyone enjoys, no matter what their age or sex. And that goes for the dogs which often steal an egg or two.

Some of the nests are never found. The first we know about them is when a broody hen walks into the yard with anything from one to fifteen chicks round her. These supply a big proportion of my replacements. The pullets I buy are R.I.R. × Light Sussex. (R.I.R. × Brown Leghorn, even from the same breeder, get disease on my farm and die, even though they thrive on some classes of land.) I keep a cock which claims a lot of Indian Game in his ancestry, and so the home-bred birds may resemble almost anything. The cockerels usually have a good wide breast, and the mongrel pullets lay quite well. At any rate we seldom go short of eggs for the house, and broodies that have reared a family usually start to lay again in late summer when eggs are scarce. We can usually find a roaster or boiler when need be.

We have even been known to have dozens of eggs for the packing station. This fills me with alarm, for it involves egg-washing and so forth, and probably means that we have too many birds eating too much food and taking too much attention.

Of course poultry keeping for a living is a different matter. Anyone who likes to buy pellets and sell eggs for a living can do so. I would almost sooner be a battery hen myself.

Broody hens in nests that we know about are apt to have a few goose or duck eggs stuffed under them. Thereafter they get scant attention, coming off to feed and fend for themselves, though duck and goose eggs get a good soaking in hot water before hatching to soften the skin inside the egg. Hens do not take baths as broody water-fowl do, and consequently do not return damp to the nests.

The cockerels from my mongrel Indian Game birds, are the only ones I fatten nowadays. When I fattened them systematically I used to get Sussex × Rhode day-olds when possible. They make superb table birds from 6–10 lb. in weight when plucked.

There is an art in fattening cockerels. It is often stated that they should only have mash. In fact they should be treated like delicate, spoilt children and given whatever tickles their fancies. Mash two or three times a day and constantly varying slightly in its ingredients makes a good basis. Boiled potatoes or swill dried off with wheat meal makes a change from the standard mash of mixed oats, wheat and barley meal. Then the proportions of the three cereals can be altered and more or less fish meal mixed in (less towards the end for fear of tainting the flesh). In the evening, when they have eaten all the mash they can swallow, they can usually be persuaded to eat a little more whole corn (barley or wheat). This lasts longer inside them and carries them on through the night. They should always have grit and water. And fresh green stuff almost every day keeps them fit and prevents their appetites from flagging. Never should they get more food than they will quickly clear up. What is more discouraging at tea time than the remains of an unfinished lunch? If they fail to eat well at any time, a day or half a day's starvation will probably do the trick. That is how I used to turn out really high quality table birds every year.

I have kept turkeys in the past. There is no trouble at all to hatch turkeys from the egg, but from then on they are born to trouble as the feathers fly upward. Turkeys wear no underclothes, which makes them the plucker's dream. But it probably also is the cause of their delicate constitutions. This delicacy, contrasting with the hardiness of goslings, made me give up turkeys, together with the fact that I dislike having to spend all my spring and summer evenings seeking turkey hens in hedges. Turkeys and Dr. Bung do not go well together.

My largest turkey weighed 21 lb. at Christmas. I took him to a rich old man who asked anxiously:

'Is it a good big bird?'

'Yes,' I said, 'it weighs twenty-one pounds.'

'Enough', he asked, 'for three hungry people?'

The bees are really my hobby, though I seldom nowadays get time to attend to them as I should like to do. The chief points about my bee-keeping, not necessarily unique, are the selection of docile, hard-working, non-swarming strains of bee, and the use of double brood chambers to allow me to neglect them in the spring and autumn when it would otherwise be necessary to super them promptly (spring) or feed (autumn). As it is, the bigger brood chamber allows plenty of room for the queen to lay and plenty of honey-storage space to make winter stores.

I gave up feeding sugar in the autumn of 1943, when I fed only a little. An immediate improvement in the health of my bees resulted. I have never fed any since then and never regretted it either. The healthier, more vigorous bees that result from natural feeding more than make up for the honey they consume. In good seasons *surplus* yields of around 1 cwt. per hive are quite frequent. Although I know this is nowhere near a record, it is fairly good for a non-sainfoin district.

Since hormone and other weed-killers became generally used in this district my bees have always got a disease in the weeks immediately following the use of the poisons by any of my neighbours. They crawl about in front of the hive and die in large numbers. The stocks get over it again later, but it is a set-back to honey-gathering.

This is just one small bit of evidence against the use of these 'harmless' substances. (Perhaps the most cogent argument against their use is that they foul the air in the country for weeks on end.)

Chalk brood, abolished since I gave up sugar, has only reappeared twice. Once during a visit of one of my hives to a very modern farm where all the latest sprays and fertilizers are used, and once here, during the period of the 'crawling disease'. It disappeared later even though the queen was the same.

Of course, though my bees are a hobby and I rely on them chiefly to supply my own family with honey, they also serve to pollinate my garden fruit crops and any clover seed crops I may be saving.

Incidentally my honey, produced as much of it is from organically manured land (including the trees and hedges on much land outside my own control), has a flavour which one customer after another has declared to be superior to most other honeys. Modest success in the past at honey shows has also confirmed this.

One year I had very little white clover blossom for the bees, but a neighbour's field was white with it where a ley had been heavily dosed with artificial manures. The clover honey that year was not as good as usual. It looked all right and there was lots of it, but the flavour was inferior.

A similarly inferior flavoured honey has been produced in

the sunless summer of 1954, when also few fruits or vegetables attained their full richness of flavour.

The fact that my honey is appreciably superior to most other English honeys in flavour may also be due to the natural feeding of my bees on honey instead of sugar, summer and winter. Honey is produced by bees after a certain amount of processing inside their bodies. It is not just collected by them from flowers. I see no reason therefore why the strain of bee, and the way they are reared as young bees, may not affect the quality of the honey they eventually make. There is also the point that where no sugar is fed, there is no chance of sugar syrup going up from the brood chamber into the honey super.

CHAPTER VII

MAINTAINING FERTILITY

Among organic farmers it is quite common to hear the view expressed that the subsoil supplies a vast reserve of all essential minerals, and that deep-rooting leys and the mechanical subsoiler will make all these available for crop production. And since the subsoil is inexhaustible ('an earthful' to use Faulkner's phrase) there is no need to worry about replacing these minerals. All we need worry about is maintaining the humus content of the soil.

To my mind this attitude is almost as criminal or stupid as extracting crops with fertilizers and ignoring the soil's needs for humus. It is not quite so bad, because you do safeguard the soil from erosion, and your crops are—to begin with—healthier than those grown with fertilizers. But in the long run both methods will impoverish the soil and ill health and poor yields will result.

If the minerals in the soil and subsoil *are* inexhaustible, then I am just making a fuss about nothing and we have got a splendid way of farming. Keep up the humus and sow deep-rooting crops and go on selling off your soil's minerals for ever. But are these reserves of minerals inexhaustible?

Of course they are not. They are very large on some soils and very small on others. A lot of my farm is on Oxford Clay, as I have said. Beneath the top soil lie nearly 400 ft. of impervious clay, unleached by rain, untapped by plant roots, virgin rock. The reserves of minerals in this must be inexhaustible, as far as human life is likely to be concerned. Even on poor clay, more deficient than most in phosphate, there will be enough phosphate in 400 ft. of it to grow about 100,000 crops of wheat without making any return of phosphate whatever. And this would be on poor clay. Some clay, even Oxford Clay, is richer in phosphate and would last even longer.

But how are plant roots to reach these reserves? Neither air nor water goes down into this clay below a certain depth. The tap roots of oak trees may, here and there, penetrate for several feet, but crop plants, wheat, lucerne, grass, and so on can only push their roots down as far as the subsoil is broken up either by natural or artificial means. Mechanical subsoiling at great depths isn't practicable. Cracking in very dry weather seldom goes down over 3 ft. and this is exceptionally deep. The fact that air never penetrates far into Oxford clay soils is proved by the shallowness of the layer of clay which is oxidized and thus turned from blue to yellow. A very few feet below the surface the clay is blue. Probably 4 ft. of soil and subsoil is as much as any crop roots can hope to use on this sort of land. Even if we could break up the soil below this depth it would not be possible to drain it on my farm except by pumps and dykes and sluices as the Dutch drain their below-sea-level land. For the levels are such that only a moderate fall is possible in the drainage ditches. Digging them deeper would cause the water in the Thames Valley to run back the other way and flood the lower levels of the clay you had broken up.

I have seen it stated that adequate subsoiling will make drainage unnecessary. So it does on some land. But it is no good assuming that what works on one farm will be bound to work on all others. On Oxford clay drainage must be sideways through pipes, mole-drains or water furrows. *It cannot be downwards.* The least penetration of water causes the clay to swell and makes it utterly water-tight.

On the other end of my farm the soil over gravel is about 4 ft. deep. The gravel is full of water and therefore impervious to roots. They go down and drink from it and no doubt get *some* minerals from it. But they cannot go on down, and if they did they would only find the same Oxford Clay which underlies the whole farm, only here it would be thinner—a mere 300 ft. or so.

And if you really examine any soil on any farm you will find the resources of the soil available to roots are limited. They may be very extensive—as on well-drained sands or brash—but they are seldom anywhere near unlimited, and to assume that you can always go deeper presupposes steady erosion of the top soil, which is just what we boast that organic farming will prevent.

I have a little book called *The Soils of Berkshire*, by Pizer. It gives detailed analyses of soil samples from all over the county, and Berkshire is a county with very varied rock formation and a corresponding number of differing soil types.

Some of these soils are rich in phosphate, some in potash, a few are rich in both. None suggest inexhaustible reserves. Man has lived on this earth for about a million years the scientists tell us, and some evidence suggests he may have existed for longer than that. What then are a few thousand years of human history?

We may be in for a long spell of geological peace. If this is so we shall have to go on farming the same soil for thousands of years if we are to survive. Yet in the top 2 ft. of some Oxford Clay soils—typical wheat land—there is only enough phosphate for 500 wheat crops. Even if you only take wheat or another cereal every other year you can only spin this out to 1,000 years, a trivial bit of time where human history is concerned.

Actually, long before you exhausted the soil's phosphate completely you would get yields dropping through its scarcity—as happens even now with some crops on this type of soil.

And who is to say that other minerals are more abundant than phosphates? Potash we can say is more abundant in Oxford Clay, gratifyingly so on some types of clay, but what about the trace elements? We know very little about them. But though they are only needed in traces many only exist in traces, and no one knows how long they will last.

There is one way in which fresh mineral matter is 'added' to the soil. Birds and earthworms which grind up their food in gizzards with the help of grit are constantly reducing larger, insoluble particles of rock to finer particles (witness the polished surfaces of the grit in a chicken's gizzard). This must expose for solution in the soil minerals which were previously locked up in the grit. Of course the process must be very slow, but it must act as a slight tonic to soils which contain grit particles and a big worm population. The process can hardly absolve us from the need to replace the minerals sold off the farm.

So I try, and I think every farmer should try, to replace the minerals taken from his soil by crops sold off the farm. Crops consumed on the farm do not matter. Their minerals are simply being moved about the farm, and provided reasonable care is taken to prevent wastage down the drain and so on they are not lost.

But crops sold off to the town or to other farms contain minerals which are lost. Most of those in the former find their way eventually to the sea.

It is impossible to return these minerals to the soil, so long as the townsman insists on wasting them. Attempts to salvage sewage sludge are a poor second best, for nearly all the potash which is highly soluble and some of the phosphate and trace elements will have been lost in the water from which the sludge is separated.

But it is possible to return substitutes for these minerals, at least in some cases.

Ground chalk or limestone will replace the losses of calcium and calcium carbonate and there are enormous amounts of chalk and limestone available.

Rock phosphate, basic slag and a number of good but expensive organic manures will replace the phosphate. (When phosphate such as slag is applied to grassland it is essential to graze heavily for the next season. Only in this way are the clovers encouraged and the maximum results from the slagging obtained.)

But to replace potash is a real problem to the organic farmer. The orthodox one can use one of the standard potassic manures. But these are all suspect in my view as being highly soluble and containing the acid radicles of hydrochloric or sulphuric acid in addition to potassium.

Organic manures such as guano, fish meal, slaughter-house refuse and so forth usually contain very little or no potassium.

There are organic manures containing potassium, but they are bulky or hard to come by for most farmers. Seaweed is rich in it. But how many of us are fortunate enough to be able to send a tractor and trailer down to the shore to collect it?

Straw and bracken contain potash, but if either or both are grown on your own land, they do not *import* anything to replace sales. They only move it about and make it more available to crops.

Sawdust contains potash, and where it is available is a useful source of fertility. Wood ashes, too, contain potassium, and none should be wasted. But where the wood was grown on the farm it is only moving potash around to burn timber and put the ash on the farmland.

If only we had not got this crazy, suicidal, water-born sewage system the problem of maintaining fertility would be simple. It is obvious that if nearly all our human and other organic wastes were put back on the land *as well as* our farmyard manure and composts of all available farm and garden wastes, then there would be almost no losses from the land. In fact so long as we were to go on *importing* food or feeding stuffs into the country there would be a steady *increase* in the mineral fertility of the soil. We should be putting in more than we took out.

In addition there would still be slag, rock phosphate, fish meal, seaweed, sawdust and so on to help out and add even greater fertility to our fields. We should soon have our land so fertile that we should not have to import any more basic foods, but would be able to export, say, wheat to pay for oranges and tea and things we cannot grow at home.

And our exports of manufactures, be they never so much reduced, would keep us a wealthy nation since they would no longer have to pay for our food.

This is not just an improbable fairy tale. George Henderson[1] and his brother began farming on a poverty-stricken bit of Cotswold soil and 'imported' food for their stock. Over the

[1] *The Farming Ladder*, Faber and Faber. O.P.

years they have so built up the fertility of the soil that fertilizers no longer give an increase, and there is no point in using them any more.

Now the Hendersons' method of farming is only possible for a few farms. For we cannot *all* import feeding stuffs from elsewhere. Someone has to grow them! But it does prove, what every farmer knows, that importations of food are a sure way to increase fertility, provided the resulting animal manure is properly used.

This island could have the most unimaginably fertile soil in the world if we would bring our heads down from the clouds of delicacy and sanitation to everyday reality. But what hope is there for a people who are too nice-minded to call even a water-closet a W.C. but now call it a toilet, or a cloakroom or a lavatory. These three words mean, I believe, a dressing-table (French), a room where cloaks may be left and a place where you may wash. It is worth noting that the W.C. and fertilizer farming are products only of the declining period of Western civilization. They played no part in its rise but are playing a major part in its collapse. It is also pertinent to ask how far the fertilizer industry could cope with the demand if 1,300,000,000 Asiatics were suddenly to become delicate-minded and install universal, indoor sanitation.

Now I do not suggest that every farmer should keep a strict account of the minerals leaving his land and those coming on to it. But it is a good idea to keep a rough check on it.

My sales, apart from a few eggs, consist of livestock, milk, wheat, potatoes. Clover or grass seed I sell occasionally, but the tonnage is small and I have no figures for the mineral content of the seeds. Hay I have only sold twice since I started farming.

My sales of pigs are presumably balanced by the mineral-protein supplements I buy to ensure adequate minerals for their growth. In fact the swill (with fish and meat) and the fish-meal or whatnot that I import for the pigs probably replace more minerals than the pigs remove when they are sold, enough, I estimate, to balance wheat sales in most seasons.

The amount of potash and phosphate sold in potatoes must

easily be balanced by the sawdust I fetch from two local sawmills.

There remain sales of milk and cattle, including calves. The potash in the milk is, I reckon, more or less balanced by the potash in the straw that I buy or beg from various sources, and the amount of potash in animals' bodies is not large.

The phosphate in the milk and the bones of cattle sold is partly balanced by that in straw, and the rest more than made up by bought phosphatic manures such as slag, rock phosphate and steamed-bone flour.

The amount of chalk or limestone imported vastly exceeds the amount of calcium sold in milk and bones. But here it is necessary to replace calcium carbonate leached out of the soil by rain which is probably far more than losses from crop sales.

Trace elements are almost certainly all right, because nearly all the manures I use are impure and unrefined, and must each contain one or more trace elements.

So the fertility of the farm as a whole is being more or less maintained as far as minerals go, and more than maintained as regards humus and availability of the minerals. But this is only possible in my case because other farmers leave me to collect swill, straw, sawdust. If everyone was trying to replace his mineral losses without using orthodox fertilizers there wouldn't be nearly enough to go round.

The remedy is not to use the fertilizers but to use the dung, urine, etc., of the town population.

It will be noticed that I have not mentioned nitrogen. This, together with water and carbon dioxide, the only remaining foods of plants, is really inexhaustible.

The air is composed of nitrogen and oxygen and traces of other gases, including carbon dioxide. As fast as plants and animals use up these gases they are replaced in the air by the natural cycles of decay and re-gassification, and by the breathing of plants and animals and by burning of fuels and eruption of volcanoes, etc. And as the air is constantly on the move over the face of the earth, every farmer gets a lavish supply of these essential plant foods.

Most types of crop cannot use the nitrogen in the air in the inert, gaseous state. But leguminous plants (clover, beans, lupins, etc.), can use atmospheric nitrogen present in the air spaces in the soil and 'fix' it in usable form in the nodules of their roots. The actual work is, of course, done by special bacteria in these nodules.

In addition there are other natural sources of nitrogen for the soil. The amount that falls as nitric acid in rain following thunderstorms is said to be very small indeed, only a few pounds per acre per year. But the amount that a healthy soil can 'fix' seems to be very large.

Scientists, working usually on exhausted soils, such as the plot at Rothamstead which has had no manure at all on it for more than a hundred years, declare that the amount of nitrogen that a soil can fix is very small.

If this is so I just don't know where my crops get their nitrogen. In ten years I have only used about ½ ton of bag nitrogen. (That was mostly for a crop of wheat which got waterlogged in winter owing to the neglect of a main drainage ditch by the local drainage authority. The rest was for experiments.) I have bought a bit of fish manure or fish meal for food, but that has not affected all the fields on the farm. Yet my crops get steadily better, taking the rough trend over the years. I never have a really poor-coloured corn crop. Even the barley I grew in 1953, the third straw crop, was *too* well supplied with nitrogen and was a very lush, dark green, right up to the time it came out in ear and began to ripen off.

Where does the nitrogen come from? Some comes from dung and urine, some from clovers, etc., some from thunder rain, and some from nitrogen-fixation in the ground. Since most of the dung and urine come from home-grown feeding stuffs, apart from the small percentage in imported straw, we may really say that nearly all the nitrogen from my crops and grass comes ultimately from legumes and fixation by bacteria in the soil. There is no other known major source.

If the scientists would test the amount of nitrogen fixed in some of my fields compared with that fixed in the most impoverished land at an experimental station, I think they might have a surprise.

In any case, wherever the nitrogen comes from, it comes, and in sufficient quantity to make up for the inevitable losses in soil and from manure, and the same has been proved by other farmers, some far more highly scientific than I am, who have given up using artificial nitrogen for a long period.

One thing I am pretty sure of is that a soil will do what is expected of it where nitrogen is concerned. If it is dosed annually with bag nitrogen it will rely on this source. If it is left to shift for itself for supplies of nitrogen but is kept supplied with humus and the minerals (including calcium carbonate) needed for fertility, it will steadily improve in its efforts to supply its own nitrogen, until quite amazing quantities are produced.

Soil analysis at Haughley has shown (*Mother Earth*, April 1955) 'that some fields in the organic section (which receives no inorganic nitrogen) showed levels' (of nitrate nitrogen) 'far above those on other sections where nitrogenous fertilizer had been added'.

This will come as no surprise to biologically-minded farmers. All the same it is good to have this sort of thing in cold figures for the benefit of the infidel.

I have used rock phosphate with good and asting results on my land, perhaps longer lasting but less spectacular results than slag gives.

But slag is wonderful stuff. Its spectacular achievements need not make us suspect its virtue. Compost and farmyard manure both give spectacular results. It is not only things like sulphate of ammonia that do so.

My friend Newman Turner[1] has waged war on slag, blaming the slagging of Goosegreen Farm by his predecessor for the ill-health of his dairy herd in his early years there.

But I am sure it was lime that began to ruin Goosegreen. In the days before the Nazi war there was a subsidy on slag, but a bigger one on lime, and the lime used in those unenlightened

[1] *Fertility Farming, O.P., Fertility Pastures,* The Rateavers.

days was usually hydrated or burnt lime. Very large amounts often accompanied moderately heavy doses of slag on poor pasture. The harmless and beneficial chalk and ground limestone were seldom used and were regarded as poor, slow-acting substitutes for proper lime.

Now hydrated or burnt lime are apt to fix trace elements in unusable forms. For example manganese can be so fixed and its unavailability can be disastrous to the fertility of a dairy herd.

I have used a lot of slag at different times. The dairy herd have relished the resulting improved sward and have milked well on it. And I have *not* suffered from breeding troubles.

Slag seems to increase the earthworm population. It certainly puts more nitrogen into the ground (through the clover it encourages) and results in more dung and more plant matter per acre.

Fungi, too, thrive on slagged land. I don't think it affects them one way or the other except by making more stock food which results in more dung and urine. I have seen it stated that all fungi, including ordinary mushrooms, disappear for good from slagged fields. On my land this is not true. I have picked mushrooms very soon after slagging a field and also many years after. The same applies to other fungi, edible and inedible.

I know such a defence of slag marks me out in some quarters for crucifixion. But I refuse to call everything Devil's Dust just because it comes out of a bag. Slag is rock which has been baked in an oven with the addition of a little lime and the loss of iron. I see nothing in this which condemns slag *a priori.* It has been said that heating ruins the 'naturalness' of anything. If so soils derived from igneous rocks are suspect. And all rock minerals were igneous once! What is more, the temperature of igneous rock was once far higher than any blast furnace. So in theory and in practice I believe in slag.

So much for the mineral balance sheet which the soil chemists talk about and which has been so unjustly ridiculed by some organic enthusiasts. What about the dung and compost which, together with humus made by sheet composting (see Chapter IX) maintain and improve the health and texture of my soils?

I used to make quite a lot of compost when labour (e.g. P.O.W.'s) was cheaper and more abundant. But now most of the straw at Pucketty is made into dung, either by feeding it or treading it, and this applies to a large but variable tonnage of straw imported on to the farm from other farmers and from a nearby mill which uses straw and sells the waste, short material.

Even without this imported straw I should make a great deal of dung. But the extra amount means that I can be lavish in bedding the calf boxes and pig sties and cow yards, and so can make more dung with less loss of the valuable urine.

Sawdust alone makes excellent dung which breaks up quickly on grassland (there being no 'long' material in it) and it absorbs the urine better than straw. But I usually mix it with straw because we get sawdust rather spasmodically whereas straw is almost always available.

Most of the dung is cattle dung, pig dung taking next place as far as quantity goes. Of course the other livestock all help, ducks, chickens, geese, horses, all make dung either in the fields or in boxes or both, and it all helps to maintain fertility. But pigs, cows and calves supply the bulk of the manure used at Pucketty.

This is carted, usually direct to the field. I do not make dung heaps. They double the handling of the dung, which nearly doubles its cost, since labour is the chief expense in making farmyard manure. We cart dung whenever we can. An odd load may go out every now and then all the year round. But occasionally we have to make a special effort to get it out, either because it has accumulated to an alarming depth in the buildings or because a particular field needs dung before the land is ploughed.

Some years in wet harvests I have carted scores of tons of dung while waiting for weather dry enough to continue carting the corn. At such times there is usually such a flush of grass that little inconvenience will be caused by fouling one grass field with dung. And if pig dung can be put on the grass the cattle will not be so much put off grazing as if cow manure had been used. Probably the latter can be put on a fallow or on a stubble between the rows of stooks or handstacks. There is usually somewhere the dung can be put except in the early summer or late spring when hay and corn crops are uncut, pasturage is short and fallows are not yet clean enough of weeds to be covered with manure. But at this time of year you are so busy that you cannot spare the time, on a small mixed farm, to cart much dung anyway.

As to quantities applied, I have usually got so much dung to get rid of that we put it on as thick as the manure spreader will put it, driving close up to the last row each time. This works out at somewhere around 24 tons per acre, though to speak of 'tons' of dung is rather a hopeless description, since the weight varies enormously with the moisture content.

My dung spreader is supposed to hold 30 cwt. of ordinary farmyard manure when level full. We usually overfill it, and it must then hold about 2 tons of ordinary dung. Very wet sloppy dung is probably far heavier than this and very light, air-dried stuff far lighter, but the amount of humus it will yield must be much the same every time.

Sometimes I have got poultry manure from a local poultry farmer. But this, though wonderful stuff, is only obtained spasmodically.

So far I have discussed the fertility of the farm as a whole. But the fertility of each field must be watched separately.

Every good farmer does this, of course. But particular attention needs to be paid to outlying fields. Fertility tends to move towards the farm buildings, so that fields near home are almost always the best. This is so for many reasons.

It is easier to cart dung to fields near the farm buildings and yards. These tend therefore to get more than their share. Then the outlying fields tend to be the ones that get ploughed up partly because they are 'poor' (and no wonder) and partly because the grass fields near home are handier for the goings and comings of the dairy herd. Also they may have to stay in grass because they are 'passage' fields across which stock and tractors must pass to reach other land.

These home fields also tend to get more than their share of

grazing (which is less exhausting than mowing) while the further fields get mowed so that their crop is removed even when they are in grass.

The hay and corn grown on the far fields will be brought home somehow, but the dung seldom goes back to replace fertility.

Another factor in this uneven process is that many farms are on more or less hilly ground and the farm buildings are often at the bottom of the various slopes. Hence a load of corn or hay or roots comes easily home and dung or fodder goes hard back up the hill. Even soil on wheels and feet seems to come into the yard and never the other way.

Within a given field the fertility tends to move towards the farm buildings, the gate and downhill, all of which are often in the same direction. Dry dung will roll downhill, not up. Rain will wash soil or bits of grass dropped by grazing animals downhill, not up, etc.

Then if part of the field is to be mown and part grazed, the far part will, almost inevitably, be mowed. And when the whole field is grazed the cattle are apt to stand around the gateway or under the hedge to do their dung.

To counteract all this it is necessary to wage a perpetual war against the incoming tide of fertility. A real effort is needed to cart dung whenever possible to outlying fields. The home fields can be largely neglected since they get grazed and have cattle foddered on them in winter and have poultry or perhaps pigs ranging over them at all times of year.

When dung is carted to any field, concentrate *first* on dunging the far end and then the middle, leaving the headlands (where cattle shelter from the cold in winter and the sun in summer) until last or not dunging them at all. It is no good starting to dung near the gate and resolving to do the rest later. So often only part of the field gets done. So make sure of the far end first.

When fodder is being carted to cattle, at any time of year an effort should be made to take it as far as possible across the field away from the gate. The bits of straw they don't eat and the manure they drop while feeding will then muck the poorest part of the field.

Another thing is to consider really carefully whether you cannot do without that little paddock which seemed so essential for a century or so. I had a paddock, 'Pig Close', which seemed indispensable. For years the fertility went on accumulating there and I only mowed it in my first year at Pucketty, before I had any appreciable amount of livestock. In the end I ploughed it up and we got on all right without it. The track across it being fenced off with two electric wires joined by a 'bridge' wire on tall posts. When it is ploughed the worm population is seen to be truly amazing.

I have for years made some effort to 'neglect' the home fields and improve the fertility of outlying ones, and it is a source of satisfaction when discerning visitors admire the soil texture and colour of distant fields which were in a very poor state of fertility when I came here.

CHAPTER VIII

DUNG AND COMPOST

I believe well-made compost is superior to farmyard manure, and experimental evidence (Soil Association) confirms this. In the winter of 1946–7 I made a big compost heap in six sections on the lines described at the end of this chapter.

Floods when the snow melted washed right over the heap. It was in a lake of about 24 acres area and waves were to be seen breaking over the top of the heap. We carted the heap away in the spring. Nevertheless the position of the heap was still visible six years later, clearly shown by the slightly better crops where it had stood.

I doubt if farmyard manure would last quite so long though it certainly lasts three to four years.[1]

At the end of the chapter I give details of how to make compost for anyone who wishes and has time to make this wonderful stuff. But I frankly admit that I make very little nowadays.

On the other hand I make a large amount of farmyard manure. Between the middle of November 1953 and 1st May 1954 we carted out 430 tons (approximately) of farmyard manure and there was plenty more we had no time to shift.

I use a Massey-Harris muck-spreader, and apart from a milking machine and a tractor I never bought a more worthwhile machine.

On a reasonably well-stocked farm the limiting factor in dung-production is bedding. The amount of bedding may be limited in which case the amount of muck made will be limited, for a beast can only produce a certain amount of dung and

[1] That is in terms of visibly better yield. Rothamstead have detected small increases due to dung for twenty years. They have never measured the effect of compost.

urine in a day. But where a lot of straw is available the amount of muck which can be made is astonishing.

Deep litter for poultry is now all the fashion, but years ago when I had a huge stock of wheat straw I used to bed up even my few laying ducks with straw and make a few tons of dung a year in that way.

Cow yards, pigsties, calf boxes, stables, all can be *liberally* and often bedded up. Give them as much straw as will keep the *surface* always clean. Underneath the urine and dung will mix with the litter and make a muck from which nothing is lost by seeping away.

If a great amount of straw is not to be had on the farm, there may be other sources of bedding. Sawdust is excellent, and has the great advantage that it spreads easily even when applied late in spring to grass intended for mowing. It also has no weed seeds in it.

As I have said elsewhere I manage to get waste straw in large amounts from a straw mill. In emergencies I use the best of this for feed. But this is seldom necessary and most of it is used as bedding. I often also get straw for the fetching of it from one or other of my neighbours.

The disadvantages of all this are that I get more than my share of weed seeds. Docks are my worst problem. They thrive on good land in good heart and are the hardest weed of all to eradicate without poisons or hand labour. If the worst comes to the worst I shall have to put the dung into heaps to heat up before spreading it.

There is also the risk that I may (I almost certainly do) get straw which has been sprayed with one or other of the poisons used for weed killing. I hope that the benefits in increased supplies of humus outweigh the dangers. So far I seem to be right in this. Some at least of these poisons are probably fairly quickly destroyed in the soil and the dosage must be very small in the ripe, threshed straw compared with what is put direct on to the soil of a field when a crop is sprayed. Certainly there is not enough poison in the resulting dung to check the growth of weeds, worms or fungi.

It is frequently stated that dung is so deficient in nitrogen that artificial nitrogen should be added to supplement the limited amount of this plant food in the farmyard manure.

Of course if dung has been leached by rain in the open it will be deficient in nitrogen; though the addition of even leached dung will so boost the soil's biological status that it will begin to produce nitrogen in larger amounts for the crop.

If, however, dung is carted straight from the calf boxes or pigsties to the field and is ploughed in at once, or if it is spread on grassland in showery weather, the amount of nitrogen it supplies must be considerable, judging from the crops that can be grown in this way without artificial nitrogen of any sort.

In any case it is often possible to supplement the nitrogen in the dung in one or two ways. It can be spread on a ley and ploughed in, when the nitrogen accumulated under the ley will be added to that in the dung. Or it can precede or, better, follow mustard grown on a fallow. Or it can be spread on a catch crop of trefoil or crimson clover or on a bean stubble, when the nitrogen from the legume roots will supplement that in the dung. When using dung to grow crops such as roots or kale which demand a lot of nitrogen I try to do something of this sort. There is no need to resort to artificial nitrogen, although it is much easier to do so than to use forethought and good husbandry.

Among organic producers, it is common to hear objections to ploughing in dung. I think it is better, when building up the fertility of a field, to plough twice or in some other way to mix the dung into the soil rather than leave it as a flat layer buried 4 or 5 in. deep. And it follows that the shorter the dung is the easier it is to mix with the soil.

With really fertile soils I have used old rotten dung, badly leached by rain, but very short and friable, with good results. I have also, on such soils, ploughed in long dung and left it buried. I have also dispensed with dung and grown a good crop of fodder beet without it.

But on a highly fertile soil the crop will not depend on the dung for its nourishment. The dung simply feeds the soil and it doesn't se em to matter much in such a case where it is put. A very fertilesoil will be so well aerated by organic matter and the burrows of earthworms (with which it should be almost *seething*) that it can cope with dung ploughed in, mixed in or scattered on top. I am of course talking about loams, my own soils. Light sands may be quite another matter and surface manuring may be much preferable.

Until a high degree of fertility is reached it is better to mix it with the soil. If necessary, plough it in in time for it to rot down short in the soil and then mix it in when you break up the furrows to make a tilth. In this way it will feed the soil and also the crop that is being grown on it.

Compost, though harder to make, is easier to handle, since it is so short when well made that it spreads easily and can be harrowed or cultivated into the surface soil. I would never plough compost in if I could help it.

There is not much more to say about dung, and I will go on to describe how I used to make compost on quite a big scale, and how I still make it, upon occasion, either for special crops such as potatoes, where super quality is so desirable, or when some particularly weedy material needs 'cooking' to kill the weeds.

I first started serious compost making as a means of using up straw.

In the early days of the war I had very few cattle. I had to grow the maximum possible acreage of wheat and other cereals to satisfy a War Committee which thought in terms of acreage rather than in yield or health of crop. Short of labour and power, I was seldom able to sow my whole wheat acreage under entirely satisfactory conditions. Yields were often poor. But even so, rick upon rick of straw soon appeared in various parts of the farm.

On this land, absolutely safe dry spots which *never* flood are not easily found, not that is, in places where ricks can conveniently be put and where they can easily be reached in winter by the threshing machine. Soon all available corners were occupied either by hay or straw ricks. I decided to give up about an acre of an old grass field to be a rick-yard. It was separated by a road from the rest of the field, which made access easy even in winter, and, being ridge and furrow land, it supplied plenty of dry ground for the ricks.

We began at one end, and did not use the whole acre at first. But gradually the ricks grew in numbers, until practically the whole acre was full with either corn or straw ricks in suitable places, and there were others in corners of fields and so on besides. At the peak of my surplus of straw, I must have had a carry-over of old season's straw each year of about 90 to a 100 tons.

That is a lot of straw for a small farm. In fact I was tactfully advised on more than one occasion to burn the lot and get rid of it. This I obstinately refused to do. By buying a dairy cow when I could afford it, and by rearing as many calves as possible each year, I was steadily increasing my herd. This enabled me to tread more straw into 'dung'.

I often marvel at farmers who burn whole ricks of straw and let their pigs and calves shiver on piles of wet dung or on cold concrete floors. Bedding stock up liberally with straw, so that they never begin to look dirty, makes more muck, and better muck (since the urine is absorbed) and it keeps the stock thriving in comfort.

By following this practice of using as much wheat straw as possible for bedding up pigsties and calf boxes, and by using as much as possible of the oat and barley and mixed corn straw for fodder I was still able only to check the growth of the straw rick population. I could not overtake the surplus. So I took to systematic compost-making.

Before this, fish offal I collected from the local town had been buried in dung heaps or ploughed into the ground direct. The same applied to feathers, etc. Now I decided to use them to reduce my surplus straw.

The Indore method of compost-making was adopted, as being the most common-sense and free from 'mumbo-jumbo'. Perhaps the other ways of making compost are not 'mumbo-jumbo'. I am more broad-minded now than I was in those days. But at least the Indore method was simple, it was based on sound and very old scientific principles, *and it worked.* I asked no more. Before this I had made compost only occasionally and rather badly. I now set about it in earnest. Fish and poultry and rabbit offal and dung were used as the 'animal' ingredients or rotting agents, i.e. they supplied the nitrogen and so forth needed by the heat-producing fungi and bacteria in the early stages of fermentation. Since the 'dung' came usually from calf boxes and pigsties or from under the bull or behind the cows (in those days I had no milking parlour) it already contained a large proportion of straw. So the proportion of straw used in the heaps was reduced somewhat.

Our compost varies in quality not only with the weather (which affects the moisture in the heap) but with the men or man making it. If they skimp the work and pooh-pooh the rules the result can be frightful. I find that when I attend to the making of a heap it is always a success.

I will describe a typical heap as it is made here and then discuss variations on the theme.

Wet straw forms the first layer unless brushwood or hedge trimmings are available to make a sort of airy mattress under the heap. The wetter the straw the better, preferably straw that has been lying flat on the ground in the rain in thin layers or small heaps in my perennially untidy rickyards. If it is already partly rotten it doesn't matter. This is thrown down as loosely as possible in rectangular layers about 6 in. thick and say 5 yds. long by 3 yds. wide. The exact size that is convenient will be determined by the size of vehicle that brings the loads. It is convenient to be able to bring one whole layer or two whole layers of material at a time.

Of course a compost heap may be built gradually over a period of a week or two, as material comes to hand. But it is never so satisfactory as a big heap made in one or two days as a special operation. We use both methods, and the heap specially made in one burst is always the better.

If the straw or what not is being thrown by hand in forkfuls from a trailer or cart it can be thrown on to a heap of the size I have described without the necessity for walking on it.

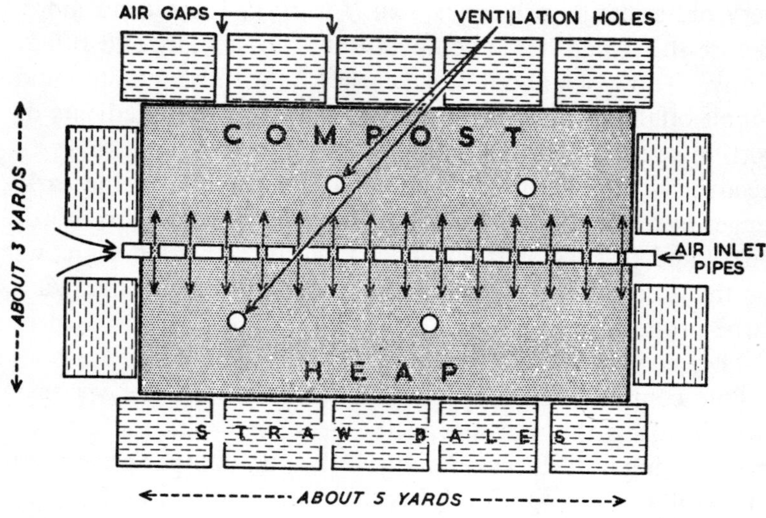

FIG. 1

Three or four poles are stood up along the middle of the heap to provide ventilation holes later.

Following the straw comes a layer of dung from the calf boxes. This is wet and rank and steaming and it stinks. Its yellow colour shows that it has had insufficient air to rot properly, for the calves had trodden it down hard. Three or 4 in. of this material may be added, for it contains a great deal of straw still unrotted, despite its content of urine. Mixed in with the 'dung' are crumbly patches of sawdust, some of them stained a bright orange, and all saturated with urine. A sprinkling of ground chalk and if possible a few shovelfuls of crumbly soil—mole heaps for example, or old ditch cleanings—are now added. About half a bucketful of chalk will probably be enough, but I believe in a good, generous sprinkling.

Next comes another 6-in. layer of wet straw and then the fish offal. The amount of this varies, according to what is available. I collect it normally twice a week. Probably not more than 1½ cwt. of fish will be available at the most. Whatever it is, it is tipped on with as little consolidation of the heap as possible. We then walk round, but not on the heap, spreading the offal uniformly over the surface, to within about 8 in. of the outside edge. If it is put right to the edge dogs and cats and foxes are apt to find it and start digging for more. If there is a whole bird, fish, rabbit or hare in the bin with the offal, it is given a fairly central position in the heap to ensure adequate rotting. Another layer of chalk follows as quickly as we can possibly manage it, as this helps to deaden the smell. A sprinkling of soil is a help again as well.

I sometimes omit soil from heaps. But generally speaking they do better for the addition of a little soil as well as the chalk. Instead of chalk, builders' lime rubble may be used, so long as it is not too full of big lumps and stones. (If it is, it should be screened.) If this has lain in a heap for some years and become a sort of soil, so much the better. Ground limestone is as good as chalk I think, but chalk happens to be more usually available here.

If a generous ration of fish was put into this layer, wet straw may follow for 5 or 6 in. But if the fish was rather scanty, about 4 in. of strawy dung will be best.

Layers of straw and dung and chalk are added until a height of about 4 to 5 ft. is reached. Then the ideal thing is to finish with a thin layer of fine soil. This is not always available, and we usually end up with a layer of dung. It does not matter if this is fresh or fairly old and rotted, though old rotten dung will tend to prevent rain from soaking into the heap. The rain is useful if the straw is not too wet when put in. The poles are now withdrawn with the minimum of treading on the heap and all is finished. Or stand on a board on the heap and poke holes with a crowbar instead of using the poles. This way makes building of the heap easier.

If the heap can be turned after about a month, outsides to middle, and water added to the dry parts, well and good. But we seldom have time for such luxuries.

Turning is not really necessary if adequate provision can be made for the air supply, and if the wind can be prevented from drying the outsides, and if the heap is made of really moist materials to start with. Really wet straw when the heap is being built will prevent any smell emerging, so that even dogs do not know what is buried in the heap. Within a few days the temperature will rise and the heap will begin to settle. Steam should pour out of the ventilation holes, and possibly out of other parts of the surface as well.

The sporophores of fungi will appear in due course and are eagerly devoured, during winter at any rate, by any ducks or chickens which find them.

I have given some thought to the provision of air to compost heaps. I think the best plan is to surround them with a containing wall of straw bales. Two bales high may be needed at first but as the heap settles the top layer can be removed for use elsewhere. The bales are not pushed very tight together, and so admit air freely between them. Then a row of land drain pipes (I have a lot of old 4-in. pipes) laid along the centre of the heap and emerging at each end allow air to enter from both open ends and percolate through the heap and out of the ventilation holes which are best staggered slightly. (See fig. 1.) If a series of heaps is to be built it is possible to run this air pipe right along the whole series, building each heap on to the end of the last, and keeping two parallel walls of straw bales to protect the flanks. I do not believe in digging the soil first. It does not seem to be necessary.

After a year or so of such use the straw bales will begin to disintegrate, and may either be added to more compost heaps or be used direct on the land as manure.

A variant of my usual plan which I once adopted when a number of heaps were to be built one after the other, in fairly quick succession, and when no bales were available, is illustrated in the accompanying sketch (fig. 2). This is a plan of the layout of six heaps. Air can of course enter all round the outside of the group, and it can also enter through the drain pipes. It can emerge through the ventilation holes. So there is plenty of air everywhere, and yet the surface exposed to the wind is greatly reduced by the arrangement of six heaps together.

Whether air did in fact enter by the pipes and leave by the ventilation holes I was at pains to investigate. A lighted match was held at both ends of the pipe. The flame was drawn in at *both* ends more or less equally (on a calm day). There was no doubt that air left by the ventilation holes. It can always be seen doing so on a cool morning, when the holes resemble a row of boiling kettles.

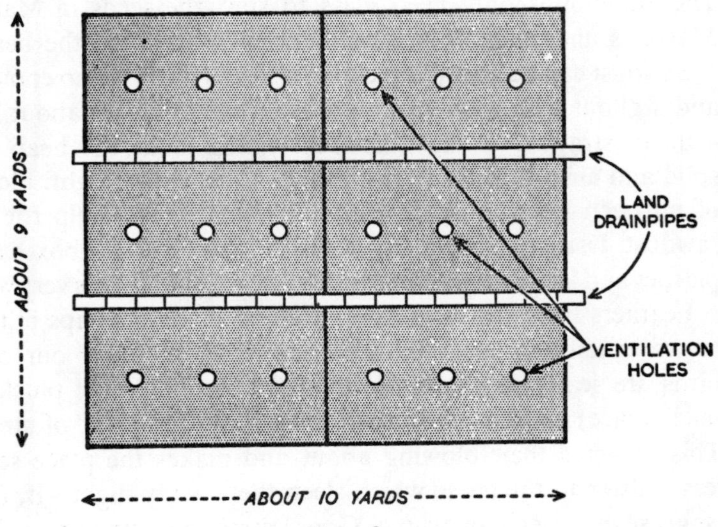

FIG. 2

That is the process which, added to the increasing consumption of straw by our growing herd, has gradually reduced our surplus straw. The acre of ricks dwindled and was cleared and the ground ploughed. In 1948 it grew a fine crop of kale, and in 1949 a splendid crop of potatoes with only a moderate dressing of compost. Good crops of potatoes were none too common in our district in that year, but these continued green and flourishing through six to seven weeks of drought when there was almost literally no rain, certainly none to soak into the soil more than a $\frac{1}{4}$ in.

Subsequent cropping of this piece of ground has included fodder beet, kale twice again and wheat. This illustrates the way fertility of a fair-sized piece of ground can be raised by making it a rick or compost yard, and then heavy cropping can be sustained year after year with moderate dressings of dung.

When a number of compost heaps have to be left during summer and have already passed the very hot stage by, say, the middle of April, they form an ideal site for marrows, ridge cucumbers and cantaloup melons, which, if a suitable strain is used will ripen quite satisfactorily in a reasonably good summer. The cantaloups only need glass to start the seeds in March. Marrows and cucumbers should be sown direct on the heaps.

Sawdust can be added direct to the heaps if it is damp enough, and if about a 1-in. layer be used at a time to replace about 3 to 4 in. of straw. Too much sawdust would make the heap too solid and airless. But mixed with the straw it is all right. Plenty of nitrogen—rich muck or fish offal—is needed to help rot the sawdust. Usually I prefer to put the sawdust into calf boxes and pigsties and let it become saturated in urine. It then rots very well.

Feathers from the fishmonger are added to the heaps in thin layers exactly as fish offal. Feathers from plucking our own birds are scattered on the floor of the stable (where plucking takes place) and covered each time with a thin layer of straw. This prevents their blowing about and makes the place seem less sinister to the geese which sleep there every night—if, that is, geese ever sleep anywhere. Goose dung, pig dung and urine, some horse and often cow dung and urine, feathers and straw and sawdust and ground chalk all get mixed together in my stable. The resulting mixture, if allowed to rot in a heap is almost as good as compost.

It is a good plan to make compost heaps in the same place several years running. Worms can then take refuge in the soil between carting one heap and beginning its successor, and can re-enter the new heap in good time. Brandlings or dung worms are very useful in a compost heap. They help to mix and aerate it. In fact, when beginning a heap or heaps in a new place it is well worth while to inoculate them with brandlings collected from old heaps elsewhere. They very soon multiply, and can then be left to invade each new heap of their own accord. Even with ordinary dung heaps it probably helps them to rot and make better manure if a few handfuls of brandlings are introduced.

If the compost heaps can be built where the dung is, or where the wet straw is, it saves labour in carting.

A point worth mentioning is the question of making compost in winter. An idea seems to have got about that compost heaps will not heat up properly in winter. I think I have read this in one of Miss Maye Bruce's writings and that is probably where the idea originated. I think it is true enough of small heaps made on a garden scale. But it is quite untrue, happily, of big heaps such as I have described. Indeed we used to make most of our best compost during winter. It is then that we have usually got more time for the job, and wet straw is always to be found somewhere. I have seen heaps steaming away furiously with snow on the ground and even on the extreme edges of the heap. But on top the snow melts and soaks into the heap. Incidentally, nature makes most of her 'compost' in autumn and winter.

When the heat has gone from a heap it will slowly freeze, in cold weather, from the outside inwards. But by that time the compost is already three-quarters made and not much harm is done.

Of course composting involves more work than ordinary dung carting, and a lot more than merely walking round setting half the countryside alight with a box of matches. But it is well worth the extra labour for crops where quality is all important or where insufficient stock exists to make enough dung. No one who has eaten compost-manured potatoes baked in their jackets, or cooked in any other way in their jackets, could ever wish to go back to the mouldy-tasting tubers of questionable interiors, sold in the shops as potatoes. And the housewife who has got used to just washing the skins of the potatoes and taking the inside for granted will not be enthusiastic about going back to potatoes which have to be peeled because the skins are uneatable, and which have to be cut open in case the inside has a brown hollow in it. Of course such potatoes are not grown with well-rotted farmyard manure, but with all the chemical nonsense of modern scientific farming.

What we want is some way of making compost mechanically

on the small farm. Mr. Friend Sykes[1] has evolved means of handling compost on his big farm, by means of expensive machinery. But the small farmer cannot possibly afford such a capital outlay as that involves.

The mechanical dung-spreading trailer is now very efficient. It may yet be improved, but I consider that it has solved the problem of transporting and spreading dung or compost. What we want now is some reasonably cheap machinery for mucking-out yards and loading the trailers. A beginning has been made at producing this sort of machinery, but in my opinion it still leaves a lot to be desired.

Where a mechanical muck-spreading trailer is not available, I usually use an ordinary tractor trailer for dung-spreading. The dung or compost is then scattered from the trailer. It is sheer waste of labour to throw the muck off carefully into heaps and then go round again scattering the heaps. With a horse and cart it is different, since the tip-up dung cart makes it very easy to drop the dung in heaps without forking it out of the cart. But with the trailer it is much easier to scatter it, either as the trailer moves slowly along, or as it stops at short intervals. Very heavy dressings of muck can be applied in this way.

Mucking out small cattle yards can be very hard work. One winter I accepted the offer of all the linseed straw belonging to a neighbouring farmer. All praise to him, he was using every scrap of wheat straw he had for a big cattle yard. But he feared the linseed would be too tough a proposition when it came to mucking it out in the spring. Rather than see it burned, and being myself, by that time, short of straw, I accepted the gift gratefully. How we were going to muck out our cattle yard in the spring I did not know, but I used the straw in the faith that we should find some way of dealing with it.

I designed a one-man-operated tractor muck drag. The idea was to back up to the dung in the yard with the drag hinged in some way on to the draw bar of the tractor, and held up in the air by the driver by means of a cord. He was then to re-

[1] *Humus and the Farmer, O.P.*, Abstract Rateavers.

lease the heavy drag which would drive its points into the dung when it fell. The points were to be such a shape that they would draw into the dung still further when pulled forwards by the tractor. The tractor, I thought, could by this means loosen up the matted straw in the yard (and even in sheds if their fronts were open) and it could then be forked into trailers with comparative ease.

When spring came we found the linseed straw in our yard was short and brittle and a girl could easily fork it up from the ground. Exactly why this was I do not know, nor whether it would always be so. But I fear my neighbour has taken to using his linseed straw, for I was honest enough to tell him about it!

So the drag never had to be made and whether it would have worked I cannot say. If the idea is of any use to anyone else, he is welcome to use it, so long as he does not claim it as his own idea and patent it, thereby restricting its use to his own advantage.

What is really wanted is for more farmers to take up compost-farming. This would bring more and fresh brains on to the job at the farm end, and it would create a demand for muck-shifting machinery which would sooner or later attract a competent supplier.

It is necessary that machinery should not only be ingenious but that it should do the job quicker and cheaper (and if possible better) than hand labour. This is not always the case with new inventions. There is nothing to be gained by watching machinery at work if you can do the job better without the machines.

I append at the end of the book a list of the materials used, when available, in compost heaps at Pucketty. The list is probably not complete even for this farm. It is all I can think of at the time of writing. For other farms, other materials would be available and the list would be different.

When I came to Pucketty I was advised to put in a W.C. instead of the outdoor E.C. I found here. This I refused to do. The W.C. will ruin Western civilization more surely than the atomic bomb, if it is not abolished, and I will not be a party to the plan. Instead I reformed the ghastly E.C. into a sanitary and wholesome affair of my own design.

All nightsoil from the E.C. is composted for use in the garden, and apart from a little straw occasionally, the garden makes no demand on the farm for its muck. Beside our kitchen sink stand two buckets, one for all edible scraps for the pigs and the other for anything suitable for compost. Into this go all the floor sweepings, tea leaves, hair-cuttings, odd ends of string, nut shells, feathers, dead flowers, old clothes and so on. When the bucket is full it is emptied on to the garden compost heap. Its contribution to the fertility of 185 acres of land may be small. But the principle is sound—save *everything* that will make compost and burn none of it unnecessarily.

A few pigs can make quite good compost. They tend to do all their dung in one corner from a very early age. This corner gets very sloppy and the rest of the straw tends to keep too dry. That is so at first. After a week or two however the damp spreads over the whole floor and fungi begin to grow in the dung. The pigs of course continue to lie on dry straw up top, while all this goes on below. The appearance of the fungi (although out of sight under layers of straw) is the signal for the pigs to go 'mushrooming'. They rout up the entire place with their snouts and devour the fungal sporophores which are forming. This mixes everything up pretty satisfactorily, and all goes well until the sty has to be mucked out to stop the pigs walking out over the doors. Then the whole process begins again. Soil gets into the pigsties on chat potatoes and on various kinds of greengrocer's refuse (collected in Faringdon) and on the roots of old lettuce and cabbage plants from the garden, and on fodder beet roots. Chalk is added as for the calf boxes. The result, together with any egg shells, cabbage stalks, big bones (out of the swill), is again a pretty good sort of compost. It gets very hot as soon as it is thrown out into a heap. Incidentally, it has to be a very big bone to survive the teeth of an old sow! They can crack in a moment a bone which would defy a large dog for hours.

Where a lot of pigs share a small sty the dung gets trodden down too hard for fungi to grow in it and the resulting dung is more like that from the calf boxes, trodden firm and airless like silage.

Even the house where the ducks spend the night can be well bedded with straw. I was going to say the house where the ducks sleep, but judging by the unclean state of ducks' eggs in the morning, they spend their nights playing football.

I fancy there is a lot of mineral matter—mud and so on—in ducks' dung. Certainly a dozen or two ducks can produce a cartload of excellent stuff about two or three times a year. Chalk is best added even to the ducks' house straw, and occasionally I throw a few handfuls of soil in there as well.

Besides compost made in heaps and farmyard manure, a sort of compost can be made by sheet-composting. Excellent results can be obtained by this method, and it has the advantage that it is a great labour saver. This is dealt with in another chapter.

Worms are wonderful composters. If a stubble is well disked so that the bits of stubble are lying about all over the place, the worms will set about collecting the stalks round the entrances to their burrows. A short time after disking, it is almost possible to count the number of earthworms in a square yard, by counting the number of little piles of stubble. Obviously you are actually only counting the big worms, for the small ones are not strong enough to collect the stalks and drag them to their holes. I once found a small fungus growing out of a worm's little 'compost heap'.

I do what every organic farmer should do—get on to the farm all the organic manure I can manage economically. On the farm scale it is economics that is the limiting factor. Each farmer must look around and see what he can find in his own district. Shavings, sawdust, tanbark, spent hops, shoddy, coffee grounds, all sorts of stuff may be available cheap or free for the fetching. Some will need composting before they can be used, and the best source for advice about this is The Soil Association, 8F, Hyde Park Mansions, London, N.W.1. They will tell you how to compost anything that will rot. But some industrial wastes don't need composting. I once got a load of cocoa-bean waste as a trial. It is not worth my while to get it regularly as it has to come too far, but it was useful stuff. It is crumbly, like very old compost, and to spread it I found the best way was to chuck a few shovelfuls on top of the dung in the dung-spreader. That spread the two together, whereas the spreader made a poor job of such powdery stuff on its own.

I use fish and poultry offal in my compost heaps because I have them. But I have made good compost without either, using any kind of dung as a starter. Now I have found a local firm with a lorry pump who may be going to make compost for me on the farm from straw and the contents of septic sewage tanks which they have pumped out.

It is always worth finding out what your local council is doing with sewage. If they are composting you are lucky. But sludge, as such, is variable, and I believe it can sometimes have industrial chemicals in it which make it a bad manure for the land. I believe Luton will supply a lime-precipitated sludge loaded free on your lorry, or spread 40 tons an acre for £6 within three miles. This sounds excellent. The trouble with all sludges is that the potash is lost in the effluent, leaving you with humus material, phosphate and nitrogen. Therefore municipal compost made with straw or dustbin refuse is better as it contains some potash. But if you can only get sludge, I would strongly advise you to get it, pocket your organic pride, and buy some potash to balance up the sludge. Otherwise you may get all straw and no grain in your corn crops.

CHAPTER IX

SHEET COMPOSTING AND CATCH CROPS

As far as materials are concerned I could probably get enough to make compost for the whole farm, every acre, every year. But the cost, chiefly in labour, would be prohibitive. Even to make and spread as much farmyard manure as that would be impossibly expensive and beyond the power of myself and three men. There are other ways of maintaining the supply of fresh humus material for the land, and I will say something about those I know about. Each farmer who works on organic lines will find out what suits him and his land best, though the season may dictate its own terms to you.

Sheet composting is a term used by Sir Albert Howard to cover any process by which raw organic material is left on the soil surface or just below it and converted to humus on the land. In short it is chucking the stuff out in a thin sheet and letting it rot. It is sometimes assumed that the only way this can be done in England is by ploughing up a ley, and since most good farmers use leys, what is all the song about sheet composting?

Leys are a very inportant source of sheet compost, but they are not the only one. There are two other main ways of sheet composting that I use at Pucketty—the catch crop method and the heavy straw method, to give them imposing terms.

A catch crop is any crop grown between two main crops, an example being the traditional stubble turnip sown on disked or quickly ploughed stubbles after an early corn crop is harvested. This can be fed off, pulled and carted off or ploughed in in time for another main crop without upsetting the rotation. It is a crop snatched as a sort of bonus on the year's cropping.

I will return to catch crops later. First let us say something about the heavy straw method. When I have a field to be fallowed the next summer, say a dirty stubble, I may decide to winter cattle on it, probably the young stock. They will get a little hay and a lot of straw and be fed on one or the other or both twice daily all winter. If the ground is wet I like to put out the straw first and lay the hay on the piles of straw. In this way no hay gets trodden into the ground as is apt to happen when it is flung down on the bare earth. In severe frost hay will be licked up clean off frozen dung pats. But in mild weather it should be put on clean ground or on straw as just described. If plenty of straw is fed the cattle can pick out the best and softest parts and tread the rest in or lie on it, which they much prefer to lying on the ground in winter.

Every day they should be fed in a different place, so that wasted straw gets left all over the field and does not go all in one place. In mild weather feed near the middle of the field, and keep the sheltered ground near the hedge for cold, windy weather. There is no need to take a tractor out to the cattle twice daily. A store of hay and straw bales or bags of chaff can be taken out in one load and this can be fed by one man on foot as long as it lasts. Do not leave the store always by the gate, if it can be avoided, or all your straw will get fed near the gate (see remarks in chapter on fertility). But a stock of bales can be left behind another hedge, in an old waggon in the field, in a portable pig house or behind the fence of a rick of hay. In this way we never take the tractor out at week-ends and holidays. My tractor driver leaves a good supply of food out in the field and I (or someone else) go out and carry the food to them on foot. Loose straw can be left on a trailer or old horse waggon in the field and the waggon can be left in a different place every time. If there are not many cattle, and no bullying, they may be left to help themselves to the straw. But small ones will get crowded out when they are feeding, then the old ones lie down on the straw they have spilled, and the young ones continue in fear and hunger. If in doubt go out with a fork, which can be left under the waggon, and spread the straw on clean ground in heaps, and make so many heaps, so well scattered that not even the timidest and smallest need go short.

Beware, too, of chaff getting in their eyes. This is particularly likely to happen with oat chaff or oat straw. Any beast with a watering eye should be brought home and examined. Probably the oat 'flier' will be found clinging flat to the eyeball. Pick it off with blunt-ended forceps and put some penicillin into the eye, the penicillin supplied for udder-injection. Make a good shot the first time with the forceps. Once the beast knows what is up it is surprising where its eye will disappear to! If the forceps fail, pour in cod liver oil and see if you can float out the offending chaff and soothe the eye. Failing this get the vet.

Sawdust can also be spread on the land and sheet-composted, but do this on a small scale, if at all, until you know what you are doing. I have done it in my garden, soaking the sawdust with urine until it goes nearly black. This ensures the rotting of the sawdust without starving the ground of nitrogen. In a very fertile paddock I have chucked down half a bag of sawdust and had it rot in incredibly quickly. But do not do this on a big scale until you are certain you know what will happen. That paddock was so rich it didn't need any more fertility. Much the safest thing to do with sawdust, unless you have a vast amount of it, is to put it in the pigsties and calf boxes and get it saturated with urine and mixed with dung. Then it can be composted or applied direct to the land without any fear that it will draw on the soil's nitrogen reserves when rotting.

When a field is thus covered with straw the dung and urine start it rotting and when the plough (or disks or what you like) goes in in the spring and the fallowing begins, the sheet composting has already started. The nitrogen fixation of the fallowing and the nitrogen released by killing weeds and bacteria and so on will supply all that is needed to complete the rotting of the straw. There is no need to fear nitrogen robbery in this way. What is more, there will soon be such an excess of nitrogen that mustard or some other crop can be sown to use it up.

Another form of sheet composting is the ploughing-in in autumn of mangold or sugar beet tops. Here again the new fodder beets have an advantage over mangolds in that the tops can be fed if they are wanted. Also there are more of them per acre than with mangolds, which probably is connected with the greater amount of sugar they manage to make in the same time.

To return to catch crops. They are one of the easiest ways of maintaining the organic fuel which nitrogen-fixing bacteria use. So do not fear that catch-cropping demands extra nitrogen from the bag or the dung cart, as some people believe. The reverse is true. The reverse is even truer if the catch crop is some sort of legume. Even plain straw ploughed in year after year without nitrogenous fertilizer has been shown, at Woburn, to give crop increases, on poor sand, of about 14 per cent. At first there was a drop in yield due to nitrogen robbery by the straw. Then the rotted straw so improved the biological status of the soil that future additions of straw rotted without nitrogen robbery. Where a green crop is used instead of straw, and where the soil has already reached a moderate degree of fertility, there will be no trouble from nitrogen shortage. Of course I am speaking about my own land all the time. I do not want to generalize about soils I know nothing about or climates in other parts of the world.

A catch crop is used to increase the supply of food or of humus or to keep the soil covered between two main crops. Anything on a stock farm which increases the food supply will add to the humus and fertility stores, but a catch crop may be grown for ploughing-in (sheet composting) without being fed off. In this case it makes a much more *uniform* addition to the humus in the soil. Provided the soil is fertile and in good biological shape it is probably better for the field to plough in a catch crop without grazing it off first. Where the crop is grazed by cattle or horses, dung and urine fall in widely separated spots all over the field, leaving nothing much in between.[1] And it is not a lot of help trying to spread the droppings since you cannot spread the urine. All the same, if the soil needs a tonic biologically, grazing is probably to be recommended, since the dung and urine patches act as centres of infection for bacteria and fungi. Sheep give more uniform distribution of manure than

[1] In the case of leys grazed several times a year the dung pats will be much closer.

cattle, which is one reason why they were so good in the old days as consumers of folded crops on the downs in preparation for corn, particularly when the corn was to be malting barley, where uniformity of growth is so important.

You cannot always do what you want with catch crops. On my land it may be too dry and hard or too wet to do any sort of cultivations at the right time, or simply too dry to make seed germinate. In 1954 I sowed mustard in early September on two fields or rather one and a half. Rain prevented our finishing the job on the second field until it was too late. September is really far too late. I normally do not sow mustard after the middle of August and that is sometimes too late in a cold, dry time. But in 1954 the gamble came off and I got lush crops of mustard knee high by November. The rain continued, however, and it was quite impossible to plough that heavy land. Rather than see all my mustard destroyed by frost I put the young cattle on to it, and they ate the lot and were glad of the green food, though it is not usually considered much use except for sheep.

Mustard is the supreme crop for another of the functions of catch cropping. I mean catching soil nitrates before they leach away with the winter rain; other plant foods, too, but they are less likely to be lost. Locking the nitrogen up in organic form in the mustard (or other crop) at least delays and reduces the loss. This catching of nitrogen is the chief reason why wheat does so well after mustard. By early spring when the wheat really needs the nitrogen, which was there in such abundance in the autumn, the mustard is rotting well and yielding up the nitrogen it saved in its tissues.

If nitrogen saving is the chief consideration, and one crop your immediate horizon, plough (or disk) your mustard in while it is still young and proteinous, just before it flowers. If, however, you want the maximum amount of humus as a soil improver leave the mustard till it is flowering well and is getting a tough, woody stem. If you want to realize how very little humus is made from young, succulent mustard in the high-protein stage try chain-harrowing it in front of the plough on a frosty morning. It is almost as disappointing as chain-harrowing soap bubbles would be. Chain-harrowing, incidentally, is the best way I know of making a neat job of ploughing it in. You run the harrow in 'lands' to fit the ploughing you will do presently.

Any quick-growing, nitrogen-demanding crop will do as a nitrate-snatcher instead of mustard. The scientists tell us that there is no loss of nitrogen by leaching on grassland. This suggests that grasses are more efficient than most crops at preventing nitrogen loss. I suppose mustard would be all right if it were frost-hardy and grew all winter, for I imagine that one reason for the efficiency of grass at this game is that it will grow in mild weather even in winter. Supreme at this mild weather growing is Italian ryegrass. It is perpetually making demands on soil nitrogen, therefore it must save a lot from loss by leaching. How thick it has to be to save all, or whether it can save all, I do not know. But it is a good idea to sow Italian ryegrass on fallows that are to be left till spring before they are planted. And you may be glad to graze it in winter or spring before you plough it in. I certainly would do so. The worst advice I ever had as a dairy farmer was to avoid Italian ryegrass as it 'took too much out of the soil'. This is negative advice. What are we farming for if not to get as much out of the soil as we can? Some of us don't care whether we ever put an adequate amount back, but we all want to get the maximum out. And for a few months Italian ryegrass is exceedingly productive. It grows in all but the coldest winter weather when most grasses are no more than slightly active if they are even green. If not wanted for the cows, a fortnight's grazing about Christmas time for the young stock is a wonderful tonic which sets them up for the hard weather which may start in January. It is also a pleasant saving of hay and straw, for on good Italian at that time of year cattle will disdain all but a mouthful of good hay, grazing with earnest concentration as if they can hardly believe their good fortune and are afraid they will wake up and find it was all a dream.

A catch crop of Italian ryegrass and trefoil (about 10–14 lb. Italian ryegrass and 3–4 lb. trefoil per acre) sown under a corn crop or on a fallow or even on an early broken-up stubble will help appreciably to shorten the winter and will make quite a sward to be turned in before the next main crop. Or the rye-grass and trefoil may be left for hay in an emergency or grazed on and off as long as it is productive and then the land can be ploughed and bastard fallowed.

A word of warning about undersowing corn crops. In a dry time you bless yourself for being so far-sighted. The ground at harvest time is like concrete and sowing impossible, and feed is getting short on the pastures with little aftermath on the meadows.

But in a wet season you are faced with sheaves whose lower portions consist of the lushest, wettest grass you ever met. The corn hardly ever dries out enough in the bottoms to let you cut it, and once it is cut you have to resort to tiny hand stacks or tripods to get the stuff dry, and it goes on raining, and the grass, instead of making hay in the sheaves is apt to make mould. And it goes on growing around the bases of the sheaves and keeps them wet. I don't want to sound too depressing. But if you live in the west or north don't go and undersow *every* corn crop until you have had a real beast of a wet season and know you can cope. In the east I imagine it pays to undersow or sow at corn-seeding time and get the catch crop established against almost certain drought. In Berkshire we are neither east nor west, nor anything (which probably explains why we seldom find a weather forecast to fit us), and I compromise by under-sowing some crops and not all. Barley is the most likely to suffer from too much grass in a wet time. Wheat and oats may smother it a bit better. Sowing after the corn is established needs good timing, but helps to check excessive growth of the grass.

In cases where the catch crop is likely to be left for hay and then the aftermath grazed before ploughing, or where several spring and summer grazings are likely, it is a good plan to include in the seed mixture even a few ounces of quick-growing white clover (e.g. New Zealand) per acre. It is surprising what a help this is to augment the trefoil both as feed and then as green manure.

In 1953 I undersowed some late barley with about 12 lb. Italian ryegrass and 3–4 lb. of trefoil and threw in as well a small amount of old white clover seed. The barley was the third unmanured corn crop and you might think nitrogen would be lacking for the early bite I hoped to get.

I grazed in March, and twice in April and then got a very good crop of hay (nearly 3 tons to the acre) followed by a certain amount of grazing in the late summer and autumn. By the end of November there was a lot of white clover as well as trefoil in the sward.

In contrast a nearby farm had a large notice board on the side of the road advertising their early bite with 4 cwt. (!) of nitrochalk per acre. The land is light and well-drained and might be expected to promote early growth much better than my heavy soil. They actually began grazing in May.

Admittedly it was a very late season. But had it been an early season I should probably have had my first grazing in February.

What is the use of such a heavy dose of nitrogen if in fact the early bite arrives late?

The only thing I dislike about Italian ryegrass is the effect on the health of cattle if they are getting little else but Italian ryegrass. Just plain Italian ryegrass by itself is a poorly balanced diet. Cattle need herbs or other food as well. If I can manage it so, I keep a lot of rough grass all winter on a good permanent pasture, or ley with herbs on it, or even a bit of rough grass of a different species. This foggage grass, being partly dead, helps to keep the cattle from scouring too much on the lush Italian ryegrass, and it supplies a useful supplement to their diet. Hay is helpful too, in the early spring, especially if it contains herbs.

Later in the spring other pastures will be available and can be run with the Italian ryegrass, or if not run with it they will be grazed one after another, and so the deficiencies of pure Italian ryegrass or Italian ryegrass and trefoil will not have long to take effect before the cattle come on to the next field.

But to return to sheet composting. The efficient orthodox farmer seldom relies upon the residual fertility of his fields, but doses the land with artificials for each successive crop. In the same way I like to see some humus-forming material going into every arable field every year if it can be arranged. It may be no more than a naturally grassy stubble after a ploughed-up ley, or it may be a coat of dung or compost or a crop of chickweed, which, incidentally, grows during the winter almost as well as Italian ryegrass and is therefore probably as efficient as a catcher of nitrates. But if the stubble is likely to be clean and free from grass or weeds a catch crop offers the means of ploughing in some fresh humus-forming material.

Where red clover seed has been saved and cleaned before being sold, the cleanings are often a useful way of getting a catch crop to plough in. Get them analysed by the National Seed Testing Station at Cambridge (if the uncleaned sample was analysed this is good enough). If the list of weeds among the clover seed is not serious, sow the cleanings with your spring corn and you will get quite a lot of clover in the stubble. The weed list may look formidable. But get a cheap 'Flora' (botanical text book) and look up the Latin names. Very likely the weeds will turn out to be ryegrass and ribgrass and similar harmless or useful plants. Of course if there are harmful weeds such as dock or dodder, the cleanings should not be sown.

The sowing of trefoil or trefoil and Italian ryegrass under barley crops is well known as a means of maintaining fertility by stockless farming, together with the use of artificials.

A field on a neighbouring farm grows trefoil naturally and the seed sheds out in the stubble under every corn crop. Before the present farmer took over, that field was always put into corn every year. I myself saw twelve successive corn crops sown there, and I don't know how many went before that! No dung or artificial was ever used, and the crops, though not heavy were not too bad. It seemed a miracle to me—until I learned about the trefoil.

Both trefoil and Italian ryegrass are very quick to seed themselves. For example, trefoil will usually have a few seeds ripe from the earliest flowers before the crop is at its maximum for hay. In this way a catch crop or ley containing trefoil or ryegrass often leaves seed in the soil which may give another excellent plant in the stubble of the succeeding corn crop. On some land, where trefoil grows naturally, slagging the arable land will induce a smother of trefoil to appear under corn crops.

The winter corn I sowed in 'Dry Meadow' in 1953 was well undersown in the spring of 1954 by plants of trefoil and ryegrass which volunteered from the previous hay crops and I have two potential catch crops or leys without buying or sowing any seed. One at least I shall leave for hay in 1955. This is an example of how it pays to avoid a strict rotation, and with benefit to the land.

I have never used kale, rape or turnips deliberately as crops for ploughing in, though I have ploughed in late turnips which failed to grow big enough to use, and, many years ago now, a kale crop that was spoiled by the fly. Nowadays, if I get a crop of, say, rape, I feed it off. Winter greenstuff is never so plentiful that you can be sure of not wanting it all before the spring. This is especially so where the wood pigeons may eat scores of tons of kale even on a small farm. But I know one man who farms in a big way who frequently uses kale, or kale and turnips as green manure, sometimes ploughing in two successive crops of kale in the same field in the one season. Kale about a foot high is suitable for ploughing in. If you let it get too tall it will get hard in the stalk and I imagine that must make consolidation of the land difficult. Alternatively, it might result in the unfortunate situation where you had to cart off some of your green crop to make drilling easier.

Any animal manure or old compost that can be spread on the land for the green crop before it is grown, before it is ploughed in or after it has gone under will add to the fertility of the field and will tone up the fungi and bacteria and worms that have got to do your sheet composting for you.

The seeding rate for mustard is usually given as 20 lb. of seed per acre. I don't think you can cut down much on this. If the crop is to be sown on land that is already fertile, and is to be left to grow tall and branching, you might get away with 15 lb. seed per acre. But until you have grown mustard and know what it does on your land in wet and dry seasons it would be safer to stick to the full rate of seeding. Especially is this so where the seed is to be sown by hand by someone who is not very expert at the job. Mustard seed is fairly easy to sow by hand being of a good size and weight and scattering well.

Rape is usually sown at about 5 lb. when drilled and 15 lb. broadcast. But the seed is smaller than mustard, and I think 12 lb., efficiently broadcast, should be enough. In the light of experience you could experiment with less. It is quite a good idea to sow some Italian ryegrass with the rape or kale when these are broadcast. It helps to fill any gaps in the plant of the brassica crop. Say 5 lb. per acre, mixed with about 10 lb. of rape or kale, should work fairly well. It will make good green manure and better feed than the brassica crop alone, and if the land does not tread too badly when it is being fed off there is the chance of a fair amount of second growth from the grass.

It sometimes happens that charlock will germinate thickly on a fallow in June or July. Where the land is reasonably clean by that time of year it is sometimes possible to leave it for a while until it starts to flower and so get a free crop of green manure. I don't know if any figures exist for a comparison in efficiency between mustard and charlock as manure. The charlock will not grow so tall, but it may grow a tougher stem early, and so make up in humus-forming material what it lacks in bulk. In any case a free crop, offering itself at a busy time of year is worth considering. On no account delay ploughing or disking once the flower appears. The first seeds will be ripening before the last flowers fall.

I have had very little experience with crimson clover as a catch crop (Trifolium incarnatum). It is said to be good. In this district some trouble was experienced with digestive disorders of cattle eating this clover, and as I like dual-purpose catch crops (feed or muck) I have rather shied off it. Trefoil is so simple, cheap and useful that I have been rather unenterprising about trying anything else. To get the best out of trefoil, as with almost all legumes, see that the soil has plenty of lime and phosphate.

Keep your eyes and ears open for odd, local sources of sheet compostable material. Your local council might deliver leaves swept up off the roads. You have only to look at what happens on grass land under trees, where the worms in autumn deal with quite a layer of leaves and don't let them smother the grass, to see that you could spread dead leaves on grass or arable (not on very young corn or leys) and let the worms compost them. The leaves need not go in a heap first.

Or perhaps there are pea or bean pods from a local cannery. Being leguminous these will rot quickly on the ground without any 'starter'. In fact I have sometimes put leguminous straw (clover or beans after threshing) on the land as a top dressing and let it rot. In this way you can even sheet compost a ley while it is still growing. Don't put it too thick. A thin coat of straw will shelter the ley and bring it on earlier than usual, as you can see by doing a little piece and leaving the rest.

CHAPTER X

ON PLOUGHING AND OTHER CULTIVATIONS

An American, Edward Faulkner, set the cat among the pigeons a few years ago by writing *Plowman's Folly*. He has written *Ploughing in Prejudices* and other works, since, in the same vein. All this need have caused less excitement than it did, if it had been more widely realized that Faulkner did not condemn shallow ploughing—especially where twice ploughing is practised.

The objections usually raised against deep ploughing are roughly these:

1. Organic matter is buried where it cannot get enough air to rot properly.
2. At this depth organic matter forms a sponge to attract water and so keeps too dry the surface soil where the crop's feeding roots are.
3. It brings anaerobic bacteria to the surface and buries the aerobic, so that both kinds of bacteria are pushed out of their proper habitat.
4. It inverts (and therefore possibly upsets) the mycelium of soil fungi.
5. Apart from burying organic matter it creates a break in the soil structure which impedes the upward flow of soil moisture by capillarity.
6. It breaks up the soil structure and so impedes rather than helps drainage and aeration. (So we are told.)
7. Regular ploughing at the same depth, or ploughing in wet weather on sticky soils (especially with heavy wheeled tractors) tends to make a 'plough pan' under the furrows, through which air, water and roots find it more or less difficult to penetrate.

8. It may help erosion by breaking up the soil's cohesion and also by supplying furrows, down which water can run to start a gulley.

9. By causing the drying or partial drying up of the surface soil (see 1, 2 and 5, above) it may interfere with the life of the soil population in the vital upper few inches.

Now all this may be a fair enough theoretical criticism of deep ploughing, and let me say I have hardly any experience of really deep ploughing. But it does not seem to apply in practice to shallow ploughing. In the case of (1) where the land is ploughed twice or thorough subsequent cultivations mix the organic matter into the surface soil, this criticism is invalid.

This also rules out (2).

(3) This may be true. But it is less so with shallow ploughing, and in any case dead bacteria are an excellent fertilizer, and the surviving bacteria can breed a new population quicker than other organisms.

(4) I wonder if anyone has ever inverted soil fungi and definitely proved that they suffered from it. Have they? Or have we just been told that it *ought* to upset the fungi?

(5) Rothamstead declare that capillarity has no place in the the rise of soil moisture. I and nine-tenths of practical men just don't believe them. But if ploughing is shallow, capillarity can raise the water to within a few inches of the surface, and the break at plough depth, if it is really operative at all, is no more than a deep mulch.

(6) Where the soil has a good crumb structure this just isn't true. Actually it helps aeration on my heavy land.

(7) Subsoiling with machines or with roots and worms will correct any fault from ploughing always at the same depth. In any case I believe a lot of panning is due not to the plough but to excessive use of the wrong fertilizers.

As for ploughing when the land is too wet: of course we all know this is bad. It does not condemn normal ploughing any more than an occasional dipsomaniac condemns the moderate drinker.

(8) This is probably partly true. But experience in countries

which have suffered from erosion clearly shows that no gully erosion need be feared where contour ploughing is used or where plenty of organic matter and crop roots maintain soil structure.

(9) The surface soil will dry out in dry weather if there is a crop removing its moisture whether it has been ploughed or not. The soil population is quite able to survive such seasonal changes.

Those who accuse ploughing of disturbing the soil too violently usually advocate severe disking which is fully as bad or worse, or else rotary hoeing which disturbs the soil far more than half a dozen ploughings.

My heavy land will set like concrete in dry weather, especially if it has first been well compacted by the treading of livestock. After such conditions ten strokes with the disks will fail to make even a moderately good tilth. In some seasons even a plough will not penetrate the soil until plenty of rain has fallen.

Yet on a friend's farm with light sandy loam I have seen a perfect tilth made by once disking after oats and vetches cut for silage.

So though I believe others on different soil can get along without the plough I know I cannot. And I get a little bored by some of the no-ploughing, no-digging enthusiasts who think that what applies to their land must apply to everyone else's.

I have tried no digging in my garden, merely loosening the surface 3 in. with fork or hoe. It works well enough. On a farm scale it is sometimes possible to make a tilth with the cultivator and harrows or cultivator and disks. And then ploughing may be avoided.

But if several cultivations are needed to take the place of one ploughing there is no saving of either time or money. And unless 'face' also has to be saved, there is nothing to gain. If you can use a short cut to a good tilth, either because your soil is in exceptionally good order or because the weather has been unusually kind, do so. You save on tractor fuel and you save time and labour. I cultivate much as an ordinary farmer does and snatch any little savings like this that I can. It is often worth trying a different implement. There are times when one stroke of the heavy drags will do as much work as two with the disks and so on. Experiment will tell you more than any theory.

I have tried mustard after the plough following oats and vetches, and mustard after several heavy diskings on the same field. There was no comparison between the results. After the plough, where air had got right into the soil for 3 or 4 in., the mustard grew lush and waist high. After the discs alone there was only a poor tilth. The mustard was stunted and poor-looking and began to flower when less than a foot in height, thin and spindly.

One time when the plough is probably best left at home is when you are trying to make a tilth quickly in the spring on heavy land which is sticky just below the surface.

But of course the real answer is that in most seasons, autumn or winter ploughing would have let the frost in and made a much better job than any tools can do in the spring. Where time is no object the plough can be used, leaving alternate rain and sun and further cultivations to make a tilth.

We might have rather more faith in the traditions of some 4,000 years or more of agricultural practice. Men have been ploughing (shallow) for about that long or longer. Some, the Chinese, are still ploughing the same land, though with lighter ploughs than we use. So we needn't be terribly scared because a few theorists or a few practical men on very light land suddenly start a rumour that ploughing is a fatal error. Forty centuries will find out most fatal errors in human practice.

There is one other point in this connection. Some no-diggers-no-ploughers rely on fantastically heavy dressings of compost, of the order of 100 tons per acre or more. This may work excellently well. But it is hardly an economical way for the ordinary farmer to use his organic wastes. If he can put 10 or 20 tons of muck to the acre he is doing pretty well. It is no good telling him he is a fool to plough and all he needs to do is to apply five times as much muck per acre. This advice is tantamount to telling him to abandon the cultivation of the greater part of his farm altogether.

25 and 26. A good crop of Atle spring wheat, most of which was carried without being stooked.

27. Cutting a fine crop of Welcome wheat.

28. Welcome wheat safe in stack.

One of my favourite cultivating implements is a set of drag harrows. The difficulty is to buy them heavy enough. Even when a blacksmith makes them to order it is hard to persuade him to put enough iron into them. I much prefer the straight-tined drag to the duck's-footed type. It is superb for pulling out squitch, which rises to the top of the tines where it can usually be cleaned off without lifting the drags. This is useful where a set of heavy harrows is hung on behind the drags to do two cultivations in the time of one, and with only one passage of the heavy tractor wheels. On heavy land in spring you have always to consider whether the wheels are doing more harm than the implement behind does good. It is too easy to destroy a tilth on heavy land by too hasty cultivations.

CHAPTER XI

HARVESTING

Corn in the east of England, grass and green crops in the west; that is the general picture of English agriculture, and all because of the weather. Most important of all is the weather at harvest time. Here in Berkshire we are between the two extremes, and it is not surprising therefore that we grow more corn than the West Country and more green and forage crops than East Anglia.

Obviously anything which can be done to make harvesting weather-proof is going to be important. I have described my methods of haymaking. For harvesting I have also copied the well-tried methods of those who have wetter climates to cope with.

An Irish tractor driver taught me how to make 'hand stacks'. These are, I believe, very much like old English 'mows'—a word which survives familiarly in the 'Barley Mow' of inn signs.

The idea is to stack the corn straight behind the binder if possible, in small round stacks, each holding approximately forty sheaves. In 'catchy' weather it is as well to cut only what can be stacked pretty quickly, and it is an advantage to cut only on one side of the field (unless a strong labour force is available for stacking) so that a broad rather than a narrow belt of sheaves is available. This saves carrying.

The labour of carrying the sheaves to make the stacks is more than that with stooking, and is no doubt one reason why the system is not more popular. There is however the advantage that when it comes to harvesting the stacks can be swept with a hay sweep (backwards if the ground is soft, and held on with a rope) to the rick; even if you collect with a trailer or horse and cart in the usual way, your carrying has been done once and for all when the stacks were built, and there is no running about after the stooks. Each stack supplies you with perhaps a quarter of a wagon load of corn, and you only have four moves of the vehicle used, and no walking to do, except between the stacks. This means much quicker and easier clearance of the field, and makes up for the slower work at cutting time.

In a dry season the advantages are not very large, though the corn and straw, being protected from the sun, are brighter and make better fodder. There is also the absence of worry. It is wonderful to feel that even if the weather breaks all the corn in this or that field is safe—if necessary for many weeks.

In a wet season the advantages are incalculable. The damage done in seasons like 1946 and 1948, 1951 and 1954 to corn in the stook was heartbreaking. It was 1946 that drove me to try something better than stooking. We put up some wheat in hand-stacks, under my tractor-driver's tuition, in August. We carted it after more rain deluges in December, threshed it in December and found the grain hard and dry and as near perfect as anyone could wish. That experiment converted me to handstacking.

In 1947 the mixed corn straw and oat straw which we hand-stacked was far brighter and more acceptable to stock than that which was picked up straight off the ground without stooking, although there was not a drop of rain the whole of harvest time.

I will try and describe the building of a handstack, though it must really be demonstrated, in actual practice, for the technique to be properly grasped. It is easy to learn the general idea, but the real refinements of how to deal with every kind of corn under all sorts of conditions come only with a lifetime of practice, and I am only a beginner myself.

To start with, four sheaves are stood up in a sort of stook. The others are laid against these, heads inwards and upwards, more are laid on these and so on until the stack's bottom is marked out. All the heads are upwards and each new layer of sheaves protects the heads of the last layer. The whole thing is just like a miniature rick in principle, but much steeper in build. The aim is to produce a sharply conical stack with the upper layers of sheaves approaching the vertical. To this end each succeeding layer of sheaves is stood more upright than the previous layer.

When this conical condition is reached, the top layer of four sheaves is tied together, using part of each head and twisting it up with part of its neighbour or merely tucking part of each head through the bond of the next sheaf. (I should say that although the stack normally begins with four sheaves and ends with four, many more than four sheaves are needed to go round the middle regions, where a sort of upwards spiral is pursued.)

With wheat, tying the heads of the last layer of sheaves causes some loss of grain and it is better to use a twine which can be threaded through every other sheaf and pulled tight and tied with a half bow.

The stack is now a neat tapering cone of sheaves, broad at the base and as pointed as your skill can make it at the top. It is ready for its cap. With wheat this is normally made of four sheaves turned upside down and arranged so that the heads (pointing downwards and outwards now) cover the gaps between the butts of the last four sheaves. To tie the cap together two people are a distinct advantage, though not absolutely essential. Numbering the sheaves of the cap 1, 2, 3, 4 in order, one person stands opposite, say, one, while the other stands opposite him, the other side of the stack, by number three. Each man then takes a part of sheaf two by the ears of corn, and likewise a part of sheaf four. Then they pull simultaneously and tie the parts of sheaves two and four together over sheaves one and three respectively. The knot is similar to that used in tying a sheaf of corn in the old-fashioned way before binders and string came in. This draws the four sheaves of the cap tightly together with their butts to the sky protecting the heads of the last four sheaves of the stack proper. Their heads, hanging downwards, offer very little lodgement for rooks, pigeons or water, and can be spread out to cover the upper sheaves of the stack in a sort of miniature thatch.

In very wet, warm weather which continues dull and misty or actually raining for days on end, some of the corn in the heads of the four cap sheaves may grow. But such losses are unlikely in any but the wettest weather, and in any case are a very small proportion of the whole crop. The losses in stooks can, as most farmers know to their cost, be very high indeed in prolonged wet weather.

If the corn is at all inclined to be green when cut it can be stacked more loosely and in thinner, smaller stacks. Wheat which is very dry can be put into much larger stacks of about sixty to eighty sheaves. But this involves carrying two small ladders around for capping, and is not worth the extra trouble in my opinion.

Actually, unless it is wanted for seed, or for some other reason has to be threshed early, it is usually a waste of time to stook or handstack wheat. If it is even moderately dry and free from green stuff in the sheaves it is best put straight into the rick. Even when the weather does not allow you to leave the wheat lying for two or three days behind the binder it can still be safely put into rick, as a rule, and will dry out in time to be threshed in the following spring. In this way much labour is avoided and often wheat is saved which would suffer loss or damage if it was stooked. I have seen crops of clean wheat stooked in perfect weather (when they could have been ricked in superb condition) only to lie rotting for the next two months after the weather has broken.

In seasons when the ground is already wet at cutting time there is a tendency for the central sheaves of a handstack, which are entirely covered by the others of the base and do not reach the outside, to get damp. The solution is to build lighter, smaller stacks with as much air space between the four central sheaves as possible.

Though the tripod method of harvesting may be superior to handstacking in wet seasons, no kind of tripod which comes to any height inside the stack can hold as much corn as a solid handstack of the same size. And provided the corn is not damp with rain or dew or very green when stacked, handstacking answers very well, and is much easier to do than building a stack around a tripod.

The handstacks they build in Pembrokeshire and other parts of Wales, and Proctor tripods, would all be improved by having a cap of sheaves put on upside-down as here described.

When the corn is once in handstacks it is safe from all but mice for, if need be, several months. Mice do not really do much damage until late autumn, by which time it is probably possible to shift all the corn, even in the most disastrously wet season. When rain interrupts carting, stooks will take perhaps another whole day to dry out. But a few minutes is all that the handstacks want and carting can go on again. This is a great advantage when corn is being threshed straight from the field without ricking it, as rain need not hold up threshing operations any longer than it would if a rick were being threshed.

I hope the illustrations make clear any deficiencies in my description of handstacking. But let me repeat that the art can only be learned by practice and example. Half the battle is the way the sheaves of each layer are set down on their predecessors, for this determines the graceful or otherwise shape of the stack as much as anything. If the corn is too long in the straw it makes handstacking difficult but not impossible. With long sheaves fewer can be put in each stack unless it is to be very big and awkward and too tall for capping without ladders.

Oats can be capped with only two sheaves as a rule, for they make a more graceful, tapering stack than wheat.

Barley and dredge corn make the dumpiest and least shapely stacks, but even so weather-proof stacks can be built with them.

Nowadays I hardly ever use strings in my handstacks. I have described the proper, Irish way of doing it. But much quicker and less perfect methods will soon be learned by experience and experiment. For example, an awkward stack can be capped with three sheaves instead of two or with five instead of four.

Very small stacks can be made where there is much green grass in the butts of the sheaves. About sixteen to eighteen sheaves can be built up, keeping the original shape of the first four sheaves. This leaves air spaces at the four corners and works quite well.

Handstacking might have saved a lot of corn in 1954 when the ground was often too wet to allow combine harvesters to operate.

Handstacks, like haycocks, can be moved about with a tractor haysweep when they have settled, which means that a young ley under the corn can be saved from destruction where the stacks stand, if they have to remain for long. With stooking this is really impossible as it is never satisfactory to move stooks. The sheaves don't go together properly the second time.

Even if the stacks are left and the clover killed under them the damage is less than with stooks, for though the individual stacks may kill more than individual stooks they are so much less numerous that the total area destroyed is much smaller.

In very dry seasons, like harvest 1955, barley can often be cut and picked up at once, like wheat, for barley is not cut till it is riper than oats. Oats, however, always need to be left in the field for a while, and in such seasons of settled fine weather I just have them chucked up into rough-and-ready handstacks. The looser they are put in the quicker they dry out, and a roughly built handstack without its cap will turn a shower of rain and keep the sun from spoiling the straw. There is no need to take as much trouble as in wet or changeable weather.

For wheat I only recommend handstacks where the sheaves have to stay in the field a long time to dry out and must be protected from birds and wet or where threshing straight from the field is intended, any time before Christmas, to save rick-building and thatching. If thatching straw is to be got from the wheat, handstacking might well pay in any season; I mean house-thatching straw which commands a fair price. It is said that organically grown wheat straw is better for thatching, and lasts longer on a roof. I can well believe this. Straw grown with artificial nitrogen often fails to last even until harvest and is attacked by fungi which cause lodging while it is still standing. It is very rare for me to have any wheat down nowadays. This is partly due to my using strong-strawed varieties and partly due to the better standing power of organically grown crops even when very heavy.

There are other advantages of using binder and thresher. Your investment in machinery and other equipment is lower. You hire a thresher, which is not as hard as it was when combines were a rarity. You don't need so much storage space, nor grain driers, nor do you have to market all your grain at harvest time. As combines become more and more popular it seems likely that the price incentive to keep wheat in rick until late in the season will increase. Probably, too, it is true that oat straw makes better feed when it is freshly threshed. The Scots used to thresh, I believe, every fortnight or so and reckoned their oat straw was much better and sweeter that way.

For stock farming, where oat straw forms part of the feed, the combine is no tool for the oats harvest. Where the straw is not wanted for stock feed, and where big acreages justify heavy capital investment, it may be the perfect machine. But I have almost no personal experience with it. It would not suit my conditions where my organically grown straw, even the wheat straw to some extent, forms an important part of my winter fodder supplies.

CHAPTER XII

DISEASES, PESTS AND WEEDS

The orthodox farmer may expect me to say that I have no diseases among my stock because I am an organic farmer. But I have too much regard for the truth and sense of responsibility to try and persuade anyone else to risk their stock on the basis of experience of mine that may be due to luck, the absence of infection or to some factor whose existence I have never suspected.

My cattle are pretty healthy, but I should hate to expose them to foot-and-mouth which neither they nor their ancestors for many generations have had any chance to 'get used to'. The history of such human diseases as syphilis confirms this view. 'New' diseases are usually deadly.

Myxomatosis is a case in point. I never expected my wild rabbits to resist it, and sure enough, my farm was one of the first in this immediate locality where the rabbits succumbed, 99 per cent of them, at any rate.

I have known people who, when they are ill—even slightly ill—retired promptly to bed and fasted for a day or so till they got better. This is all very fine in bed. But what farmer can just go to bed for two days whenever he has a cold? And if you don't go to bed, but keep on working, you have to eat or collapse.

Much the same goes for cows. A brief fast is all very well, but you can't expect a cow to milk on an empty belly. The best way to fast a calf (with e.g. white scour) is to give it warm water and gradually give it stronger and stronger milk and water to get it back on to whole milk again. That is a treatment I do approve of and have used repeatedly.

With mastitis (from which my cows usually suffer very little) I simply use penicillin. It is quick, harmless and easy—as 'natural' a medicine as most 'natural' remedies. The best plan with all diseases (including mastitis) is not to get them. The way to avoid mastitis is to avoid bought food and feed something green (or failing that something succulent—like mangolds) every day of the year. (Disinfecting everything is next to useless.) For four years I hardly had any mastitis. Lately we have had an outbreak owing, I believe, to bought foods (when my corn was still awaiting the threshing tackle). Previous to those four years I had an outbreak every spring from about February or March to April or May.

It occurred to me that this was after the kale was finished and the cows had been getting only hay and corn (all dry food) from about January onwards. The arrival of spring grass soon put paid to the mastitis. So I took to growing more kale and Italian ryegrass and cereals for spring grazing and shutting up grass fields in late summer to make foggage, and so on. Result: no mastitis except following severe udder injuries. Sometimes a mastitis clot will show in the fore milk. But the trouble usually clears up without penicillin treatment. Experience since these words were written confirms in general that faulty diet is the cause of mastitis. 1955 showed that even grass is not good enough food when it is scorched up by excessive heat and drought. One case I had responded to penicillin treatment only when the herd moved into a field of fairly succulent red clover aftermath (third growth after two hay cuts).

Ringworm on calves usually does not cause trouble where silage or greenfood is fed regularly. The best combination, perhaps, is kale and fodder beet. And it normally disappears soon after they have gone out to grass in spring, but not always. In some seasons it seems to hang on for months.

The only other cattle disease I am concerned with is Johne's Disease. (Abortion I never have. I use S.19 vaccine on most of my heifers and find it entirely satisfactory.) I regard Johne's Disease as a disease of dirty water (the vet says it is a disease of pastures) and I have only had one case since we had water troughs in each field and that case may have been due to not drinking at the troughs but at a pond.

I gave that case the full nature-cure treatment. After a week I had to call in the knackers. That sounds pretty damning for nature-cure. But one swallow does not make a summer. I may have been late starting the treatment or some other factor may have come into it—there may, for instance, have been some other disease or physical weakness that I did not know about. The claims of nature-cure practitioners should be more thoroughly investigated by qualified vets. Some of the cures of Johne's disease and other things that we hear about may be cures of diseases with similar symptoms, or they may be only temporary cures. But it is not fair to dismiss all such cases as bogus without thorough examination of the evidence.

Tuberculosis is a disease of buildings. Cattle reared outdoors seldom if ever suffer from it. Probably the fact that my cows lie out all the year round is responsible for my freedom from T.B. as much as my methods of crop husbandry and feeding.

But if the buildings ever had heavy infestations of T.B. germs, they failed to affect my calves. Only one home-reared calf of mine has ever reacted to the tuberculin test since I began to 'go T.T.'. That calf was one of a family which produced other obvious 'screws' in my pre-testing days—a clear case of hereditary weakness.

Organically grown food may have something to do with this resistance to T.B. but, as I say, it may equally be the open-air life the cattle lead from a few months old, together with good luck in having the infection die out in the calf boxes before I began rearing calves in them.

I am wrong. There is one other disease which afflicts my stock, though seldom severely, and that is husk. Usually a little coughing is all we get. Leys are less likely to give trouble than old infected pastures.

In 1954 the sunless summer seemed to affect the health of cattle and human beings alike, also crops. And we have had more trouble than usual. However it is rare to be forced to bring a beast indoors on account of husk alone. In this I am probably lucky rather than clever.

Pests

The subject of pests is one of the most disputed items in the whole controversy between organic and orthodox farmers.

I divide pests into two classes, the higher and the lower animals. This isn't theory. It is just plain observation over a period of years.

The higher animals include such things as rooks, jackdaws, pigeons, rabbits, rats, and so on; the lower ones such creatures as slugs, flea-beetles, wireworm, etc.

Roughly speaking I find that the higher animals delight in the good flavour of my organically grown crops and are often a perfect nuisance, seeming to come from miles around to devour them. I am sure in 1953–4 we got *all* the Scandinavian wood pigeons! But the pests in the other group seem to *dislike* my organically grown crops and leave them alone.

Now I quite sympathize with the sceptical reader who up till now has thought me a fairly moderate crank and not as mad as some organic enthusiasts—and who now says, 'Well, really, there are limits to what I can believe!'

I agree that this view of pests is utterly improbable and cannot be justified in theory—not yet. But it is simply the conclusion I have come to as a result of observation. I can't explain it, but I have observed it.

Rats, rooks, pigeons, mice, voles, rabbits, pheasants, squirrels, all these seem to enjoy organically grown food, and sometimes to show a marked preference for it. The same goes for livestock though they usually have less choice in the matter.

But I have dug up slugs sitting on compost-grown potatoes without so much as nibbling the skin. Flea-beetle seem to leave healthy, organically grown kale crops almost untouched. Snails and slugs will not, as a rule, eat lettuces in my garden. (1954 was the exceptional year in this as in many other ways.) Nor do slugs touch my wheat crops.

With wireworm I am not so sure. When I first ploughed up my old permanent pastures they were riddled with wireworm and I suffered severe losses of cereals through them. The land was far from being in a high state of health and fertility then, and the old, tough turf made adequate consolidation difficult. So I will only say that wireworm no longer troubles me now.

My observations of slugs and snails are perhaps worth mentioning in more detail. I have noticed that they only seem to attack our lettuces while they are wilted following transplanting. If they are thinned instead of being transplanted they hardly ever get attacked. If they are planted in pockets of damp compost they usually survive for they root again so quickly that they are only wilted for a very short time, and that mostly during the daytime when slugs and snails are not much abroad.

I have seen slugs very numerous indeed on a field of wheat following broken-up red clover where a lot of clover stalks and roots were lying about on the surface. They apparently ate dead clover and did not touch the wheat.

When it rains after hay has been mown and half made, especially on the sides of the roads, slugs and snails can be seen in large numbers feeding, not upon the live green grass, but upon the half-made hay, and rooks can be seen on the hayfields at such times. I suppose they are eating slugs and snails, for they are not eating the hay!

If you see a slug on a healthy dandelion, e.g. one growing beside a country road, it will be eating, not the young flowers, nor the leaves but the dead, brown end of the old flower which is still clinging to the half-ripe, unopened seed head.

Pull up a dock or pick a leaf of hogweed and throw it down on the path of a garden infested with slugs and snails (as ours is) and after dark on a summer night you will find slugs and snails in plenty have come to eat it. Yet they don't eat the live dock or hogweed.

To protect from slugs young cucumber plants before they are established pick dock leaves, etc., and scatter them around the plants as food for the slugs. The cucumbers will often survive in this way, although slugs are very partial to them.

What does all this amount to? Simply this: a slug or snail

will never eat a healthy plant if it can get a sickly, dying or dead one. Even compost appears to be a preferable food for slugs rather than the potatoes growing in it.

This doesn't 'explain' anything, and I can scarcely hope that the sceptic will believe it unless he actually observes it for himself. If he grows his crops on poison sprays and fertilizers and they suffer from slugs and snails and all the rest, he will just dismiss my claim as ridiculous, as I admit, it seems.[1]

It has been suggested that organically grown plants may produce chemicals in their tissues which are unpalatable or poisonous to insect and similar pests. But I have seen very little experimental evidence to show whether this is or is not the case.

Another possible clue to the problem is the undoubted fact that the lower animals can survive on a much more limited diet than the higher animals. And they may thrive on foods which are poisonous to us. An example is slugs eating poisonous mushrooms and apparently liking it. (Do they perhaps die later on in their beds in agonies?) Conversely, disinfectants harmless to us may kill germs or small parasites.

It is interesting in this connection to note that small boys and others who collect caterpillars can rear perfectly healthy and vigorous butterflies and moths even though they feed them partly on wilted food or on a supply of their food plant picked and kept in water.

So perhaps the slugs, caterpillars and so forth can manage very well on the poor stuff grown with chemicals, whereas we know that such food has an inferior flavour which explains why the higher animals don't like it.

But this does not explain why the slugs avoid the organically grown plant except in so far as we have seen they have preference for dying or sickly food.

Another possible explanation of at least some of these observed facts is that a caterpillar which feeds on a compost-

[1] Perhaps the orthodox farmer will be more willing to believe my claims if he considers lice on cattle. Cattle in poor condition often suffer from lice, but the lice disappear after the cattle have had spring grass for a few weeks.

grown kale plant is so tasty that no bird can resist it and it has a correspondingly small chance of surviving.

I can only sum up by repeating that I don't know why. I only know what. Probably there is no single, simple explanation covering all the facts.

Resistance of certain strains or varieties almost certainly comes into it. For example apples grafted on one stock will be healthy, while the same apple on a different stock will be unhealthy on the same soil.

Then again charlock is attacked by the turnip flea-beetle. But I doubt if it is ever wiped out by it even on soils which are far from healthy by the standards of organic farming.

Generally speaking then, the lower-order pests prefer unhealthy plants and the higher-order ones prefer healthy ones. An apparent exception is the wasp. No matter how healthy fruit trees may be, nor how splendidly produced the honey, the wasp will try and have her share and yours too. Why these insects should rank as 'higher animals' where taste in food is concerned I don't pretend to know, but it seems it is so. Perhaps the fact that they are predatory has something to do with it.

Whether caterpillars on kale, etc. are to be classed as disliking healthy plants I am not quite sure. I never seem to suffer from them on the kale to any extent. But that may be due to the activities of small birds—sparrows, warblers, etc.—which eat the eggs and caterpillars and not to any intrinsic virtue in the kale plants. Yet if caterpillars do attack the kale it is never all over, but only an odd plant here and there.

This usefulness of sparrows in one way when they are such a nuisance on half-ripe corn crops near hedges and buildings is typical of many 'pests'. Rooks undoubtedly do good as well as harm, so do jackdaws, crows, jays and magpies. Moles are a curse in a field of root seedlings but do a lot of good in helping to drain heavy land, especially as they tend to work to and from the ditches.

The more I see of the way the different forms of life tie-up with one another and react upon each other the less certain I become of where I stand with any of them. Probably it is best to let well alone. Only when serious trouble comes is it advisable to interfere with wild life.

If foxes are taking chickens, then trap or poison or shoot them. If rooks are spoiling a piece of corn, scare them and shoot them, otherwise leave them alone.

Two animals I must say a good word for are stoats and weasels. They are both excellent killers of rabbits, rats and mice. If I get a stoat taking up residence in my corn ricks I am delighted. I know my corn will be practically free from rat and mouse damage. I have never had to shoot a stoat or weasel. I have only known one case of a stoat attacking my poultry in fifteen years, and they will have to do much worse than that before I forget the ricks of mouse-free, sweet-smelling corn they have given me even when we did not thresh until March or April. Stoats often get blamed for poultry which rats have killed. Stoats also seem to eat a lot of wood-pigeons, though how they manage to catch them I cannot say.

One result of the myxomatosis which I foresaw was the drastic reduction in numbers of rats and mice following the disappearance of the rabbits. Stoats, weasels, owls and foxes have all had to rely more than usual on rats and mice for their food. (Foxes seem to have managed well enough on my poultry.) So I refrained from poisoning rats in the autumn. I hope this has enabled a fair number of stoats to survive the winter to be ready to deal with rats and mice this summer, also any rabbits that may have survived. Since the outbreak of myxomatosis a friend of mine has lost a litter of new-born pigs (outdoors) to a fox.

Owls and hawks, too, do far less damage (if any) than they are supposed to do. I never harm any of them.

Weeds

Orthodox farming tends to regard as useless or pernicious weeds many plants which organic farming regards as useful herbs. The typical example is the dandelion, of which I have written in the chapter on leys.

29. Beginning a hand-stack of wheat. Note sheaves collected in circle ready for building.

30. The finished hand stack,

31. Potatoes. Those on the right are from compost-grown Scotch seed.

32. A fine crop of healthy Gladstone potatoes grown in the drought of 1949. Bucket holds about four gallons. Heavy rain in September caused the splitting of some tubers.

But though this applies to some weeds, in my opinion a lot of sentimental rubbish has been written and talked about the virtues of weeds. Our task is as always to steer a sane course between unenlightened orthodoxy and 'organic' fanaticism.

Let me say how far I welcome weeds and use them. 'Weeds' in pasture such as dandelions, yarrow, hawkbit, ribgrass, and 'weed' grasses I welcome within reasonable limits as useful herbs. (This does not mean I welcome buttercups. Buttercups have nothing to commend them but their beauty, and rest-harrow is even worse.) Sowthistle (the small-flowered annual kind) I welcome on fallows or in silage crops where they can be used green and succulent. They have a wonderful effect on the secretion of butterfat by cows. But in other crops they are simply weeds and better absent. Even in kale they seed and die before they can be used, and the kale would have grown better without them. In corn they may be still green when it is cut. And they are so juicy and succulent that they amount to a nuisance by hindering drying of the sheaves. In hay they can be used as part of the crop when eventually they dry out, but I don't think they make very good hay as dandelions or plantains do. Even the giant plaintain, though an awful nuisance in other ways, makes a small amount of good hay, relished by stock. A little squitch in corn crops makes good 'hay' in the butts of the sheaves. The same goes for the small bindweed with pink flowers. But a little squitch this year will be a lot next year unless it is dealt with. Fortunately it is a fairly easy weed to kill given two or three hot days at the right time (i.e. when the squitch is on top of the ground). Patches of nettles make good hay (or fodder, if wilted first).

Other weeds such as docks, thistles of all kinds, fat-hen, scarlet pimpernel, charlock, coltsfoot, horse tail are just a nuisance. Perhaps dock, coltsfoot, horse tail and creeping thistle should be in a class by themselves, because ploughing them in seldom gets rid of them as it does with the others. Charlock, scarlet pimpernel, fat-hen and chickweed often make useful contributions as green manure if ploughed in before they seed.

Some organic farmers reckon to winter young stock to a

considerable extent on weed grazing. This may be possible in the warm, wet west, and no doubt it is good for the stock, but only chickweed makes any winter growth on my stubbles. Dandelions may start early in spring, but they, plantains, docks and even squitch die off during winter and sowthistles are completely killed by the first severe frost.

Self-sown cereals and grasses there may be in the stubbles, but I don't really class them as weeds. On my land a much better idea for winter grazing is a catch crop of Italian ryegrass or rape and turnips or kale. (Rape should be fed off before kale as it suffers worse from wet and frost.) Or a field of foggage can be saved when there is plenty of grass in late summer and autumn. Or wheat or other cereal can be sown early and grazed in early spring. It may even be possible to get two grazings off wheat and then a crop of corn. Grazing will usually check any tendency to grow too tall and lodge. However it is not the simple affair that some people seem to think. Grazing wheat is so tricky that only God could be quite sure when it was desirable. On my land the ground may 'tread' too much, with damage to the wheat and the backside of the man who eventually has to ride the binder. And as you never know whether you are in for a drought or the most growthy season for years you can only guess whether grazing will improve the height of the straw (i.e. shorten wheat which would have been too tall) or result in a crop which is under-strawed. And too little straw means too little corn, for the grain is made in the leaves, just as the roots are, by the action of light on chlorophyll. Then again if you graze too hard or too late you may so damage the main tillers that they pack up and leave you with only the secondary ones on each plant. That means that you have done away with what would have been the finest heads in the crop. As with most things in farming you must try for yourself and be careful, and don't think that what works one year will work the next. How easy (and how dull) farming would be if it did!

One use for weeds which we may mention is as a tonic or medicine for stock. I have already dealt with the herbs which most organic farmers deliberately include in their leys and tolerate in their old pastures. But there are other plants, many of them only obtainable in gateways or hedges, which cattle will eat from time to time. You may say that they are daft and do daft things without reason. However it is at least possible that they know instinctively what to eat for certain digestive or other disorders, and it is this assumption that herbs and weeds may supply useful remedies that lies behind much of the practice of nature-cure treatment. After seeing the things cattle will eat when they are not desperately hungry I am inclined to think that there is more to be found out about herbal remedies than the wisest witch ever knew. And remember that much priceless wisdom was probably burnt with each of those unfortunate creatures.

The following list does not mean to include every weed that any cow has ever eaten. It is just a list of weeds that I remember seeing them eat at one time or another. Many of these observations have been made when getting the cows in for milking when I have been acting as cowman. I have seen them eat charlock, moon daisy, nettles (even while uncut), flowering heads of creeping thistle, white dead nettle, stinking mayweed, giant plantain, dock, large and small bindweeds, Linaria Elatine (a species of toad-flax), Persicaria and knot-grass, ivy, elder, privet, and, of course, the leaves of such trees as the willow, lime, elm and so on. And I have seen a goat eat, with impunity, bryony (Bryonia dioica) which is poisonous. However, the way cattle will eat things like yew and get very ill or die shows that they can make mistakes.

One other 'use' for weeds is the information they may give about soil conditions.

Severely poached grassland produces giant plantain. Sour dock indicates lime deficiency. So does charlock if it is found to be suffering from club-root disease.

But many of these weed indicators are quite bogus. Moon daisies are supposed to indicate shortage of lime and heartsease (the little wild pansy) abundance of it. Which (if either) is right I don't know. But they can't *both* be right for I have found them growing with their roots tangled together.

The big 'Jack' thistle (not the creeping thistle) is supposed to indicate good land. A wise old man once said to me: 'If you are looking for a farm, look for some good old jack thistles and some big elm trees.' I would add: 'Look for strong hawthorn hedges.'

Lumping all weeds together just as 'weeds', they usually indicate poverty of soil. Foul crops are usually starved crops. The remedy is good cultivations and plenty of humus.

Hammer-mill grinding of grain will not eliminate all the weed seeds. Quite small-mesh sieves will allow whole seeds of dock, charlock or fat-hen to pass through uninjured. If the pig (or cattle) dung is not allowed to heat up, these seeds will, many of them, germinate when the dung is spread on the land. The more thorough your cultivations following an application of dung the more of these weeds will be destroyed in the seedling stage.

So much for the 'uses' of weeds. If you have them use them if you can. But try not to have them! Foul land means a lot of hard work. Weeds cannot be tolerated in crops like fodder beet. Kale will usually win through, given fertile soil, and dominate all weeds (except perhaps a bad infestation of squitch). But the weeds it eventually dominates will retard it and will seed before they give in, and it is much better to eliminate them entirely if it can be done.

One thing I am sure about. No weed is so serious that the land need be sprayed with poisons to eliminate it—at grave risk to the health and even the very lives of human beings, livestock and wild animals.

One result of the introduction of selective weed killers has been to make squitch or couch more prevalent then formerly. When land had to be cleaned of all sorts of weeds before crops could safely be sown the squitch automatically suffered with the rest. Nowadays thorough cultivations are not needed. Their effect on fertility is produced instead with fertilizers, and their cleaning effect with sprays. But squitch is such a close relative of wheat that it survives the sprays which wheat survives, and it has come into its own on some farms as the one undisputed weed of the arable land.

There are two main ways of getting rid of squitch (apart from hoeing it off in root crops, which can be very hard work and needs a hoe that is first cousin to a razor). One way, you plough fairly heavy or very heavy soils in the spring when they are still a bit sticky and let the furrows bake right through so that everything in them is killed. Squitch is seldom to be found deeper than 4 or 5 in., so this treatment can get all of it. The trouble is that on heavy land, if the baking is really successful, it may take the rest of the summer to make a tilth. In other words you are in for a whole fallow.

The other way, you plough soon enough to let the weather (preferably frost) into the furrows and then pull the squitch out of the ground with cultivators and harrows. You do this as often and thoroughly as you can and cross-plough if need be to break up the soil still more and let you get at what is still in the ground. By this way you may hope to avoid a whole fallow, and it is what I mean by a bastard fallow, i.e. one that either starts part way through the summer or stops part way through.

The first way (whole fallow) may fail if no weather arrives capable of baking the furrows. Then the only hope is to keep ploughing and wear out the weeds by repeatedly burying them. The second way will only fail if you are not thorough enough in your work or if (which is not likely) the weather is so much of the moist, growing kind that you can't kill what you pull out.

The most futile waste of time, to my mind, is to collect the squitch runners and burn them in little or great heaps. If you have got it dry enough to burn it has already said good-bye. Instead of burning stuff that can do nothing now except make humus, you do much better to spend your time and effort in pulling out the rest of the runners that are underground and which are laughing at the little fires above them. What is out is done for, what is in is unaffected by the burning.

Creeping thistle is just like squitch in that it runs underground. But it does it much deeper than squitch with the result that ploughing only cuts it off without getting under it all. To get rid of a bad infestation in arable land the only hope is to keep cutting it off either with a plough or thistle bar or with hoes. Repeated cutting will eliminate any weed by starvation. For any plant repeatedly robbed of its leaves will starve since its leaves make its food. In a garden I have even got rid of bindweed on one small plot by hoeing once a week for two years! You begin in April and take a rest in August or September when this weed normally stops growth.

A thistle bar is a steel blade that runs horizontally a few inches below the surface of the ground. It is mounted, instead of the back row of tines, on an ordinary cultivator. It cuts off or pulls out everything growing. However it is only possible to use a thistle bar when the soil is fairly dry and crumbly and free from a large amount of long rubbish or dung.

Perhaps my most serious weed at Pucketty is the dock. I get dock seed in the straw I buy for litter, and as the dock likes fertile, well-limed land it usually thrives when it gets here. It is sometimes possible to kill docks by getting them on top of the ground and exposing them to severe frost, or by either of the methods I have described for squitch. The trouble is that while two hot days will finish squitch, the dock is thicker and tougher and more cunning. It will lie on the ground and root again from any part of the root that is touching the soil, and it will go on growing upwards every time it is ploughed over. In this way you may find docks that have grown their roots in a sort of square. Very dry summers give you a chance to deal with docks. Very wet ones make pulling them easier, though it is always difficult slow work and very costly in labour.

If you have killed the dock roots with the help of a dry summer try and get seed in the ground to germinate so that you can kill the seedlings and make a job of extermination. Docks often manage to set seed even in their death agonies. If the autumn is dry and there is a risk that there is seed in the ground, you may do well to wait and sow a spring crop of some sort rather than put in autumn-sown corn only to find that the dock seedlings come up with the crop.

CHAPTER XIII

ON CHANGING OVER AND OTHER MATTERS

I may as well say straight away that I do not feel very well qualified to write on the subject of changing over to organic manuring from orthodox methods. In my own case very little change had to be made. I had never used large quantities of artificial manure. I simply gave up what I had used, including compound fish manures, containing artificials as supplements, and bought pure fish-meal and other organic manures instead as far as I could afford them. I cut out superphosphate entirely. But this was no great wrench, as I had always had a preference for slag and organic sources of phosphate.

I had never been an enthusiast for sulphate of ammonia. I stopped buying a ton or two of nitro-chalk which I had bought every year and have never regretted it.

I had always made a fair lot of dung. I increased my efforts, and made compost as well.

During the war I was not allowed to buy potash manures, as my land, being heavy, has good reserves of potash. It was, therefore, no hardship to continue doing without potash.

For farming potash-deficient land, of which I have no first-hand experience, and where 'organic potash' is not available—e.g. straw—I can offer no reliable advice. But I should be inclined to suggest a type of farming which did not involve the sale of too many products rich in potash.

Where a farmer has been using even moderate amounts of artificial manures, it will mean a drop in yields, if he suddenly stops using any at all. Probably he would be well advised to change over one field at a time, on the lines suggested by Lady

Eve Balfour in *The Living Soil*, so that at the end of a whole rotation or a period of several years his whole farm will be run on organic manures. This has the advantage that a heavy application of dung or compost can be applied to the field which is due for the change, so that the lack of artificials is less likely to be felt. But I will refer the reader to Lady Eve Balfour for very competent instructions for carrying out her method of changing over.

Remember that it takes a very long time to build up fertility in a soil to a high level over a whole farm. Anyone can make a show field by putting all the farm's supply of dung out into it year after year, until it has a tremendous amount of humus and the soil is as black as a prairie. But the aim should be to build up all the land. And since we have got to get our livings from it while this is being done it will be all the harder to do.

Nature is content to take centuries in building up the fertility of her soils, and she has no need to export all the while, as the farmer must. So do not be impatient. Remember that good results can be obtained from soils which are not yet perfect. So long as the land is improving steadily, what does it matter how long it takes to reach perfection or near perfection? And I don't think I ever heard of erosion on land which was on the upgrade, however slowly.

There is a widespread belief abroad in England to-day that nothing is worth doing unless it gives an immediate prospect of a return in money. No one will plant oak trees, not because the timber of oak is no longer saleable, but because it takes 120 years to mature. But this sort of attitude has got to stop if we are to restore our civilization to a state of health. What sort of heritage should we have in England now if our Elizabethan, Jacobean and Georgian ancestors had been content to live in 'prefabs' and plant only spruce trees and to farm as if the world were coming to an end in ten years' time?

There is only one way to farm and that is for your successors. Incidentally that is one very good reason for patriotism. I may be thought narrow-minded, but I should make small effort to farm well if I thought my land would soon be seized by foreigners.

The best way to ensure good farming is to give a farmer a vested interest in the land, and the best possible vested interest is that of hereditary ownership or hereditary tenancy. It is no coincidence that an age which has seen the most reckless farming in English history has also seen the most rapid and casual changes in the ownership of land.

I do not wish to disparage those single or childless men and women who are applying themselves with selfless devotion to the service of their particular piece of soil. But for ordinary folk it is a great stimulus to have a family interest in the land and its welfare. Our ancestors in their wisdom made even the office of king or queen hereditary, so that he or she might have the strongest of all motives for watching over the country's interests, a task which we now largely prevent the Sovereign from performing. I do not wish to discuss the whole question of the conditions for the ownership of land—and obviously there must be conditions—but I do wish to make a plea for family farms, for children in plenty to follow their parents and carry on their work; to call a halt to the 'get rich quick for to-morrow we perish' attitude of modern life, which has lately spread so disastrously into farming.

Sir Albert Howard said the land's most important crop was healthy men and women. Think then of your children and farm for them. That is the normal human thing to do, however unusual it may be to-day. Do you want your children to eat devitalized food and diluted poisons, so that they grow up as part of an emasculated race? Let no smart salesmanship or political quackery persuade you that any short-term policy is of the least use for the land. If organic manuring is the best in the long run, then stick to organic manuring.

I hope I have given the reader a fair picture of my farm and my farming. It is not the only way to farm this land, but it is a way anyone *can* farm.

I will not attempt to give costings for organic farming, because there are so many imponderable factors. To take one small example: with fertilizer farming you can assess with some accuracy the cost of your manuring and you know that it will have precious little effect after the one crop for which the manure is applied. But with a field which is dunged it is impossible for the ordinary farmer to assess accurately the money value of the dung, since he cannot analyse it, either chemically or biologically. And it is impossible to say how long the dung will go on improving his crops. Three to four years later it will still *show* to the eye. But how much increase is it giving then or in five years time? How much has it saved in vets' bills? Or in tractor fuel by making cultivations easier?

But there are obviously big savings in organic farming to set against the extra cost of extra dung-carting.

There is the saving of all expense for artificial nitrogen for a start.

Then there is a big saving in potash and phosphate. For where the soil is biologically fit phosphate and potash will be made available even when they exist in unavailable form.

So the organic farmer can reckon to use *all* the slag he applies some time, and obviously doesn't have to apply so much as the orthodox farmer who reckons that up to 90 per cent of his phosphate manures becomes permanently fixed in the soil in unavailable form.

Another example of the sort of saving that can follow organic methods is in haymaking. Soil Association and other scientists have proved that the addition of nitrogen fertilizers to grass and other crops gives an increase of crop which is at least partly due to increased water content. Grass grown organically will have an appreciably higher dry matter content than the same grass top-dressed with artificial nitrogen. When it comes to making hay of these two lots of grass the lower water content of the organically grown crop may turn the scales between success and total failure. Even if this is not the case, the organic crop can probably be made into hay quicker, with less turning and expense and with less loss of the valuable leaf. To put a money value on this sort of thing is beyond me.

I have not done any controlled experiments to test this with actual hay crops. I don't want artificially grown hay. However, casual observation of the relative rates of drying of my own and other people's hay crops suggests that this may be a factor of practical importance and not just a theoretical argument.

I cannot balance up all the financial pros and cons and prove that organic farming pays better than orthodoxy. On a short-term basis I doubt if it does so. For example, my land is not yet at the peak of fertility possible to land of that type, or very little of it is, if any. That being so I could probably increase my yield of wheat by 20 per cent by giving it an appropriate amount of artificial nitrogen. It would probably lodge (and if so I should have to buy a combine to deal with it, plus drier, etc., etc.) but assuming it did not, you could prove that it had 'paid' me to use the nitrogen bag. Gradually, however, I should find that it needed more nitrogen to get the same increase, that other crops in the rotation needed nitrogen which had not done so before, that the land was harder to plough and cultivate, that pests and disease appeared where they had not worried me before. I should be in the clutches of the well-meaning agricultural scientists, using in turn all their own remedies for their own inventions, and then the remedies for the remedies and so on like the big fleas and little fleas.

I hope no one who has read this book thinks that I fancy myself as one of the best farmers in the district. On the contrary I know that there are many cleverer and more able farmers in my own small district farming on orthodox lines. They say it is the shrewdest business men who most often fall victims to the confidence trickster. In the same way some of the best farmers fall for all the specious semi-truths of the 'experts' and salesmen, and will do the daftest things in the name of efficiency.

I should like to see some of these clever and efficient farmers (the ones whose careful plans never miscarry as mine so often do) seriously try out some of the ideas I have put forward in this book. I don't mean try to prove them wrong. You can always arrange to prove what you *want* to prove. But give them an intelligent, fair trial.

Let one of these supermen set aside a good field, give it slag (and potash if it is short of it), dung it and grow leys and catch crops and farm it well, but without sprays, insecticides or artificial nitrogen. And let him persevere even though at first his yields drop, and even if the initial lack of biological fertility results in damage from pests.

He would do everything so much better than I do that after a few years he would produce the most convincing demonstration of the soundness of organic farming: he would turn his whole farm over to it.

I do not pretend to make as much money as these supermen, many of whom farm much more land, but I certainly do not regret my own methods, even financially. My land is becoming more productive without an ever-increased outlay on ever more expensive fertilizers. I save an enormous sum on artificial nitrogen alone every year on what I would use if I farmed by the book and the fertilizer advertisements. And though I am not a wealthy man I could afford a good many television sets (assuming that I could find a use for even one) with the money I spend on school fees every year.

Perhaps it is the children that are the greatest reward of all for the organic farmer. My wife and I have bad teeth, yet the children's teeth are the delight of the dentist (in small quantities—he says he would be out of a job if everyone had teeth like theirs). It is in no sense of boasting, but in all humility that I thank God I took up organic farming early enough to produce a family that are so much stronger and healthier and more intelligent than their parents.

My belief in the exhaustibility of the subsoil does not mean that I spend a fortune on slag and potash. It merely explains why I take so much trouble to get compost material and organic manure on to the farm. And I *would* use more minerals if I thought the land needed them.

If the minerals under my land are exhaustible, how much more so are those of the natural deposits which we mine! Gone already are the superb deposits of Peruvian guano. The North African phosphate beds will go next (within two centuries, it is estimated, and in less than one if you start 'efficient', modern agriculture in the Far East). And chemical fertilizers, especially nitrogenous ones, take fuel and power to make them. (And sulphate of ammonia and superphosphate need sulphur, of which there is an acute world shortage.) We may not run right out of fertilizers, but increasing fuel and freight and labour costs keep on pushing up their prices. If any government, defying the vested interests, removed the fertilizer subsidies, so that they were sold at their real cost, then organic farming, which holds its own even against the weight of the subsidies, would be away to an easy win and those who laugh at muck and magic would find that it was muck and money. My production is based firmly on biological (or as some would say 'natural') processes, and does not rely on anything so precarious as a fertilizer subsidy. Not that I grudge chemical farming a special subsidy all to itself. It needs it, for it is like an incurable alcoholic who feels so ghastly when he is sober that the only thing to do is to have another drink.

It is significant that this subsidy, which is of inestimable value to the fertilizer manufacturers, is specifically excluded from applying to organic manures. This is war, and shows that the organic movement is beginning to worry the big vested interests, as well it may. It can't be much fun to invest millions in a thing and then have a lot of wretched little men come along and prove that it is not wanted!

In addition to the savings I have already mentioned, organic farming must involve considerable savings in vet's fees, cost of dairy replacements (I am told the average life of a dairy cow is now only four lactations—despite penicillin) and costs of spraying and dusting and disinfecting and so forth. I don't know what the sprays cost for an acre of corn, but I have seen very appreciable losses of corn resulting from their use.

On one farm near here there was, in 1954, a superb crop of wheat. In the spring it looked to be worth two of mine. It was, in fact, too good, and the farmer (rather wisely, I thought) grazed it off. The crop was making a good recovery from the grazing when he sprayed it.

All the leaves were burnt off as if a severe frost had struck it overnight. The crop was apparently destroyed. But it came again and I watched with interest.

This is the final picture: quite a good piece of wheat, but not able to hold a candle to mine which had previously looked inferior. Through the crop were lines of flattened corn where the tractor wheels had gone and where the wheat had only partly recovered. The straw was now pathetically short. And above the crop, clearly visible from the road, were docks, poppies, sowthistles and clites (goose grass). Had they developed immunity? What did he spray for? What was the final cost in labour, spray and loss of crop? No one can say. But it must be large for the loss of crop alone.

One final touch was the fact that every wheat plant grew deformed leaves. After spraying, the topmost leaves all grew twisted into corkscrew shape instead of flat. This alone must reduce efficiency of photosynthesis and result in loss of crop. And it proves that the poison affects the crop's metabolism, i.e. it gets right inside it.

My crop is not yet threshed. But it is a very good piece of wheat which no one need feel ashamed to grow. My oats, over 6 ft. high in places, are one of the finest crops of oats I have ever seen anywhere. As I write, on a wet morning, I can see them before me, golden in their beauty but with still two-thirds to cut. A difficult season has prevented our cutting more than a few rounds with the binder at the right time (a week ago now) and the straw is now giving way and more and more of the crop is going down. Even so the damage affects less than a quarter of the field. These enormously tall straws with very heavy loads of grain have stood up to appalling weather—some of the heaviest rain I ever remember with gusts of very strong wind at the same time—and even where the oats are laid they are only badly bent over. The straw has not broken and gone down flat.

I have only once put wheat on my very best land. The variety was Redman. About a third went down, the straw not being equal to the task, and the sparrows and other birds had, I estimate, about 13½ cwt. of corn off the one-acre patch. Yet we thrashed 27 cwt. of wheat. So the yield ought to have been about 40 to 45 cwt. given no sparrows and stronger straw.

I hope to try one of the newer and stronger varieties next year on the same type of land.

But the best land apart, I reckon I can now grow about 32 cwt. per acre with wheat. And I expect my yields will go on rising. I don't imagine I am anywhere near the peak yet. I am quite sure that organic methods will, in the long run, not only produce better food than orthodox farming can. They will outyield it on the same type of land.

So no one need be afraid that organic farming means lower yields and starvation. The yields may be lower to start with, just as a drug addict may be in poor shape after he has given up his drugs, or as an underfed and overtired man may be unable to do much work. But as fertility, *real* fertility comes, so yields climb and returns improve. And the improvement is real and lasting and not bought at the expense of posterity.

BIBLIOGRAPHY

This is not a catalogue of every book on Organic Husbandry. It is a severely pruned and selected list—a bare minimum for those who are interested enough to pursue the subject further. Most of the books themselves carry Bibliographies which will guide the reader's steps on again if he so desires.

D.S.

INTRODUCTORY

The Green Leaf, compiled by LOUISE E. HOWARD O.P. A selection of extracts from the writings of the late Sir Albert Howard, compiled by his wife. O.P.

Organic Husbandry, compiled by JOHN S. BLACKBURN (Biotechnic Press, 1949). A symposium. Reprinted articles covering every aspect of the subject. O.P.

The Earth's Green Carpet, LOUISE E. HOWARD (Faber and Faber, 1947). Lady Howard expresses her husband's teachings in simple language. This should be read before his books. O.P.

GENERAL—THEORETICAL AND PRACTICAL

An Agricultural Testament, SIR ALBERT HOWARD (Rodale). The story of Sir Albert's work in India on the Indore process of composting. The working out of his theories in practice.

Farming and Gardening for Health or Disease, SIR ALBERT HOWARD (Faber and Faber, 1945). This carries his thesis further. O.P.

The Living Soil, LADY EVE BALFOUR (Faber and Faber,

revised edition 1949). A fine combination of theory with practical description of work on the Soil Association's research farm. O.P.

Agriculture—A New Approach, P.H. HAINSWORTH (Rateaver). The first book from the Organic School that really sets out to play an 'away match' at the other school. It answers orthodoxy in its own language.

Fertility Farming, F. NEWMAN TURNER (Faber and Faber, 1951). Detailed practice on a 250-acre compost farm, giving costings, weekly time-tables, methods of feeding and treatment, etc. O.P.

Humus and the Farmer, FRIEND SYKES (Faber and Faber, 1946). The author practises large-scale farming on compost lines. O.P. Abstract Rateaver.

Ploughman's Folly, E. FAULKENER (University of Oklahoma Press). The case against the mould-board plough. First published in America.

A Second Look, E. FAULKENER (University of Oklahoma Press). The author has some second thoughts and answers some criticisms. O.P.

Harnessing the Earthworm, T.J. BARRETT (Faber and Faber, 1948). Describing its function in soil building and plant nutrition, and how to use and propagate it.

Compost for Garden Plot or Thousand-Acre Farm, F.H. BILLINGTON (Faber and Faber, 1942). A practical little book describing methods.O.P.

Making Friends With Your Land, LEONARD WICKENDEN (Devin-Adair Co., U.S.A., 1949). This is a book for the sceptical scientist, for it is written by one who was just that. He was converted and uses his knowledge as a chemist to explain why.

ON SOIL EROSION

Our Plundered Planet, FAIRFIELD OSBORN (Faber and Faber, 1949). The best of several books showing how we are ruining the earth's crust and must return to organic husbandry to save it.

ON OTHER CIVILIZATIONS

Farms of Forty Centuries, F.H. KING (Rodale). First published at the end of the last century, this book has been reprinted many times and still remains a classic, describing in detail the organic farming practised throughout China and Japan.

GARDENING

Organic Surface Cultivation, GERARD SMITH (Ward Lock, 1950). Compost Gardening without digging. O.P.
Gardening With Compost, F.C. KING (Faber and Faber, 1944). Covers the subject well in a small book. O.P.
Gardening Without Digging, A. GUEST (R. Wigfield and Co., 1947). A completely practical booklet on the subject.
Practical Organic Gardening, BEN EASEY (Faber and Faber). A complete text-book on organic gardening and composting methods. O.P.

PERIODICALS

The Farmer, edited by F. NEWMAN TURNER, quarterly, from Ferne Farm, Shaftsbury, Dorset. Discontinued.

MATERIALS USED IN COMPOST HEAPS AT PUCKETTY

At One Time or Another

Straw of all kinds—wheat, oats, barley, rye, beans, peas, linseed.
Dead leaves.
Mouldy hay.
Grassy strippings from ditch sides.
Sawdust, bark and rotten wood.
Squitch runners and other weeds, e.g. thistles.
Kale stems and roots, potato haulm.

Dung of all sorts—cow, pig, horse, goat, poultry.
Offal of poultry, rabbits and fish, including shell-fish, shrimps, etc.

Feathers.
Dead cats, foxes, etc.

Soil, including that from the bottom of ponds and ditches.
Lime rubble.
Ground chalk or limestone.

I have also put swill into compost heaps when I had no pigs, following swine fever. Margarine seems to be rot-proof, though it might melt in the middle of a big heap.

I once tried putting hair from a ladies' hairdresser's into my heaps. Unlike the hair of other animals it won't rot. I suppose it is so much treated with this and that as to lose its nature.

Green weeds such as nettles may improve the biology of the heap, but they are not necessary as far as heating up goes, at least on a farm-sized heap. Nor does an all straw and dung heap appear lacking in sweetness or friability when finished. The 'compost' made naturally on the floor of the wood or elsewhere will seldom contain any greenstuff. Not that that need deter us from including greenstuff if it is available.

INDEX